Science and Fiction

Editorial Board
Mark Alpert
Philip Ball
Gregory Benford
Michael Brotherton
Victor Callaghan
Amnon H Eden
Nick Kanas
Geoffrey Landis
Rudi Rucker
Dirk Schulze-Makuch
Rüdiger Vaas
Ulrich Walter
Stephen Webb

Science and Fiction – A Springer Series

This collection of entertaining and thought-provoking books will appeal equally to science buffs, scientists and science-fiction fans. It was born out of the recognition that scientific discovery and the creation of plausible fictional scenarios are often two sides of the same coin. Each relies on an understanding of the way the world works, coupled with the imaginative ability to invent new or alternative explanations - and even other worlds. Authored by practicing scientists as well as writers of hard science fiction, these books explore and exploit the borderlands between accepted science and its fictional counterpart. Uncovering mutual influences, promoting fruitful interaction, narrating and analyzing fictional scenarios, together they serve as a reaction vessel for inspired new ideas in science, technology, and beyond.

Whether fiction, fact, or forever undecidable: the Springer Series "Science and Fiction" intends to go where no one has gone before!

Its largely non-technical books take several different approaches. Journey with their authors as they

- Indulge in science speculation—describing intriguing, plausible yet unproven ideas;
- Exploit science fiction for educational purposes and as a means of promoting critical thinking;
- Explore the interplay of science and science fiction—throughout the history of the genre and looking ahead;
- Delve into related topics including, but not limited to: science as a creative process, the limits of science, interplay of literature and knowledge;
- Tell fictional short stories built around well-defined scientific ideas, with a supplement summarizing the science underlying the plot.

Readers can look forward to a broad range of topics, as intriguing as they are important. Here just a few by way of illustration:

- Time travel, superluminal travel, wormholes, teleportation
- Extraterrestrial intelligence and alien civilizations
- Artificial intelligence, planetary brains, the universe as a computer, simulated worlds
- Non-anthropocentric viewpoints
- Synthetic biology, genetic engineering, developing nanotechnologies
- Eco/infrastructure/meteorite-impact disaster scenarios
- Future scenarios, transhumanism, posthumanism, intelligence explosion
- Virtual worlds, cyberspace dramas
- Consciousness and mind manipulation

More information about this series at http://www.springer.com/series/11657

Michael Carroll

Lords of the Ice Moons

A Scientific Novel

Michael Carroll
Littleton, CO, USA

ISSN 2197-1188   ISSN 2197-1196   (electronic)
Science and Fiction
ISBN 978-3-319-98154-3   ISBN 978-3-319-98155-0   (eBook)
https://doi.org/10.1007/978-3-319-98155-0

Library of Congress Control Number: 2018954935

© Springer Nature Switzerland AG 2018
This work is subject to copyright. All rights are reserved by the Publisher, whether the whole or part of the material is concerned, specifically the rights of translation, reprinting, reuse of illustrations, recitation, broadcasting, reproduction on microfilms or in any other physical way, and transmission or information storage and retrieval, electronic adaptation, computer software, or by similar or dissimilar methodology now known or hereafter developed.
The use of general descriptive names, registered names, trademarks, service marks, etc. in this publication does not imply, even in the absence of a specific statement, that such names are exempt from the relevant protective laws and regulations and therefore free for general use.
The publisher, the authors, and the editors are safe to assume that the advice and information in this book are believed to be true and accurate at the date of publication. Neither the publisher nor the authors or the editors give a warranty, express or implied, with respect to the material contained herein or for any errors or omissions that may have been made. The publisher remains neutral with regard to jurisdictional claims in published maps and institutional affiliations.

This Springer imprint is published by the registered company Springer Nature Switzerland AG
The registered company address is: Gewerbestrasse 11, 6330 Cham, Switzerland

"Without adventure, civilization is in full decay."
                                   Alfred North Whitehead

"Everyone has secrets. Few are worth the keeping."
                                   Melciéna Valentine

"Tis true my form is something odd,
But blaming me is blaming God;
Could I create myself anew
I would not fail in pleasing you.
If I could reach from pole to pole
Or grasp the ocean with a span,
I would be measured by the soul;
The mind's the standard of the man."
                        False Greatness by Isaac Watts, 1706

Acknowledgments

Chapter 31—and, in fact, most of my writing—offers a nod to my favorite American author, Ray Bradbury. Despite its setting, a book like this never develops in a vacuum. My thanks goes to many, including Dr. John Coates, director of the Energy Biosciences Institute and senior faculty scientist at the Lawrence Berkeley National Laboratory, for his advice and insights regarding bioreactors and microbial energy, which are key points in our plot. To asteroid expert Dan Durda goes my appreciation for help on Wentaway, along with "shooting crap in a barrel" (and other carefully instigated malapropisms that helped to flesh out the character of Tony). And speaking of language, my friend Anna Petraglia Miller helped me with my Italian phrasing. Good job, Anna! Planetary Science Institute's Candy Hansen-Koharcheck provided great insights into the Enceladus Ocean and geyser plumbing. Ahoy thar, Candy! JPL's Morgan Cable reviewed my descriptions of Enceladus, a place she knows much about (I think she may have even visited). On the engineering side, Robert Zubrin helped me with details of the cycling ship, while Mike Rogers taught me all things operatic. Ben Clark's book, *Mesorg Madness*, inspired my analogy between bee colonies and certain microbes that will remain unnamed until you, dear reader, have a chance to come across them. My thanks, too, to our neighbors Jake and Mary Beth, for blessing their huge, good-natured dog "Arman." His name was perfect for one of my characters. As with so many of my projects, Marilyn Flynn was a trusted, dependable first reader and critical to this book's success, both mechanically and philosophically (with special commendation for royal etiquette). My spiritual brother Nicolas Currat helped with character development (mine and that of my novel's imaginary ones). My writing buddy Carol Eaton helped my princess along with her dialog and traditions. She and the rest of the Mile High Scribes contributed to my continuing maturation as a writer. As always, Caroline Truman Carroll, my wife, helped me to focus when I needed it most, gave me valuable life insights, and made life—and writing—just plain fun. Love you, sweetie!

Watery Critters

Several people have wondered whether my Naiads were inspired by del Toro and Taylor's wonderful *The Shape of Water*. Sadly, the answer is no. As so often happens in creative projects, my book was well into its final draft by the time this fine movie came out (writing and producing a book is a bit like having a baby: it takes at least 9 months from contract to submission.) Still, I found the film beautiful and inspiring in many ways.

A word about our hero's name: Gwen Baré is named after Jeanne Baré (or Baret), the first woman to circumnavigate the world. She was an accomplished French botanist who sailed aboard the ships *La Boudeuse* and *Étoile* from 1766 to 1769, disguised as a man. The vine bougainvillea is named after a specimen she collected in Brazil.

Contents

Part I   The Novel

1	Gwen Baré	3
2	Interval 1: Grandma Baré's Diary	9
3	Carter Rhodes-September 2229	11
4	Gwen	13
5	Rhodes	19
6	Interval 2: A Fairy Tale	29
7	Gwen	31
8	Chevalier	35
9	Gwen–October 2229	39
10	Antonio Vincenzi	47
11	Circles Within Circles	53
12	Interval 3 -Tommy- The Recent Past: 2226	61
13	Staying In Touch	65
14	Interval 4: Abandoning Enceladus, The Past: 2201	69
15	Sisters	73
16	Offering	77
17	Interval Five: Tommy	81
18	Top of the Loop	83
19	Planetary Yachting	87
20	The Water and the Wattage	95

21	Drop	99
22	You Can't Go Home Again	105
23	Down the Rabbit Hole	111
24	Gwen	119
25	The Downside of Topside	121
26	Cold Awakening	125
27	The Hosts of Atlantis West	127
28	In Trouble and Under Water	133
29	Civilized Conversation	137
30	Meetings	141
31	A New Light	147
32	Interference	153
33	Final Approaches	159
34	A Walk on the Sunny Side	165
35	The Dark Side	173
36	Goings and Comings	177
37	An FTL on the Wall	181
38	Long Distance	185
39	Taina	189
40	Homeward	191

Part II The Science Behind the Story

41	The Science Behind the Story	197
	Venus: Home sweet home?	197
	Enceladus: where the action is	199
	Energy from Microbes: a brainstorming session with microbiologist John Coates	202
	Broadcasting Bacteria	206
	Incoming!	208
	Naiad Brain Power	212

Also by Michael Carroll from Springer

Other Novels for Springer's Science and Fiction Series
On the Shores of Titan's Farthest Sea
Europa's Lost Expedition

Nonfiction
The Seventh Landing:
Going Back to the Moon, This Time to Stay

Drifting on Alien Winds:
Exploring the Skies and Weather of Other Worlds

Alien Seas:
Oceans in Space
with Rosaly Lopes and a team of talented authors

Living Among Giants:
Exploring and Settling the Outer Solar System

Picture This!
Grasping the Dimensions of Time and Space

Earths of Distant Suns:
How We Find Them, Communicate with Them, and Maybe Even Travel There

Antarctica: Earth's Own Ice World
Michael Carroll and Rosaly Lopes

A Little Ancient History

May 1887, London

 Princess Alexandra sat in her personal coach, her forehead against the glass. The locomotive rested on a track that curved in such a way that she could see the carriages behind her own.

 "Are you all right, Ma'am?" her well-coiffed attendant asked.

 "Oh, Esmie. I dread that last leg to London, the frenzied schedule that faces us."

 "Perhaps it is the cost of your generosity."

 Alexandra smiled as she pulled away from the window. "London Hospital needed the new buildings. Prince Albert and I had the wherewithal, and our advisors thought it prudent. Is the Prince on board yet?"

 "Yes, my lady. I believe he is consulting with the engineers."

 The Princess leaned forward, craning her neck to see out the window again. "Esmie, do you know if there is other royalty on this train?"

 "I do not."

 "There is a commotion back there, someone being ushered into a coach behind us. Find out for me."

 "Yes, highness." Esmie stood to her full five-foot height, unruffled her wide dress, and stepped from the cabin. Soon, she returned with a steward, who bowed low.

 "Your highness, you wished to know about the private carriage toward the back of the train?"

 "Yes, I like to keep track of my surroundings, especially if there are other royals nearby. I might like to visit."

 "Oh, I see. Yes, of course." The steward shuffled his feet and looked at the floor. "It's just that, you see, that particular carriage has not been reserved for any government official."

 Alexandra gazed at him, letting the silence simmer. "Well then, who is it for?"

 "It is a curious case, your highness. The passenger's name is Joseph Merrick. He came from a visit to the country, to the estate of Lady Knightly in Northamptonshire."

 "He must be very wealthy, to demand his own private coach," she said.

 "Hardly," the man scoffed. He got control of himself. "Excuse me, your highness. They say he is not quite human. He has the head of an elephant with huge ears and wrinkled skin. And I have heard other things as well. They are taking him back to London Hospital. That is where they keep him."

"You speak of him like an animal in a zoo," the Princess frowned.

He shrugged unapologetically.

"And what was he doing in Northamptonshire?"

The steward seemed to gain confidence then, as if he had finally been asked a question he knew something about. "They told me he spent most of his time walking the grounds and picking wildflowers. He does not stay in the house itself, of course. They keep him in the Gamekeeper's cottage."

"And what is Lady Knightley's interest in this creature?"

"I couldn't say, your highness."

Alexandra looked out the window again, lightly brushing the underside of her chin. A man in a physician's coat stepped into the private car and the door sealed behind him. "I see. Thank you. You are dismissed."

The steward left. The Princess turned to her attendant with a raised eyebrow. "Very curious, wouldn't you say, Esmie?"

***

The Prince and Princess of Wales settled into a spacious, elegant apartment with a view of St. Stephen's Tower and the other buildings of Parliament. Alexandra watched the boats float down the Thames, gulls screeching overhead. A layer of black soot blanketed the horizon. The apartment was low enough that she could hear the street vendors calling out below, selling fruit, fish, and fresh bread.

"I wonder just how fresh that bread is," she said to her husband. He was signing documents of state.

"Yes, my darling." Increasingly absent. So important.

The Attendant Royal knocked on the door.

"Come," Albert called, not looking up.

"Your highness, the carriages have arrived."

Alexandra enjoyed the short trip to London Hospital. The sounds of passersby, the smells of fresh-cut flowers and burning coal, the feel of the cobblestone avenue—it all seemed so otherworldly to her. Even the Thames smelled better these days, with all that engineering of sewers and such. At times, she wished she did not lead such a sheltered life.

The royal cavalcade was greeted by a man who looked very much like the one who had entered the mystery passenger's train car a day earlier. Their attendant announced the Prince and Princess of Wales.

"Frederick Treves at your service," said the man of medicine, bowing before the prince. "We are grateful for Your Royal Highness's generosity toward our facility."

"We look forward to seeing the new additions," the Prince said. He nodded to his attendant.

The attendant puffed up his chest and said, "May I present the royal Alexandra Caroline Marie Charlotte Louise Julia, Princess of Wales."

Treves bowed low. "Your highness."

A Little Ancient History

"The day is not getting any younger," the Prince snapped.

Treves stood and led them through the door, as he had been counseled to do by the royal couple's steward. As they walked down the corridor toward a brightly lit foyer, the Princess said, "I would like to make a request."

The doctor stiffened and paused, turning. "Anything, your highness."

"I would like to meet this John Merrick."

"Joseph," he said reflexively, a look of horror washing over his face. He recovered quickly. "He lives in the basement of the hospital, in personal quarters. He is badly deformed and generally keeps to himself."

"Is he chained?" she asked.

The doctor straightened his shoulders and said, "I have found Joseph Merrick to be an intelligent, sensitive man. He is more civil than many of my colleagues."

"Would you ask him if he would accept a visit from us, after the dedication, of course?"

Treves looked at the Prince. The Royal seemed flustered. The Prince shot back, "We are on a very tight schedule, Alix."

"It will not take long to dedicate our buildings if you aren't long-winded, and I think our luncheon can wait just a little. We breakfasted late, and I have heard so much about the Elephant Man." She turned to Treves. "He was part of a show of curiosities, was he not?"

"He was. But the carnivals enable people who are less fortunate to survive, to earn a living."

"Sad," she murmured. Meeting Treve's eyes, she asked, "Is his face truly that of an elephant?"

"No, your highness." Treves' voice took on the tenor of patience. He had answered this question before. "Joseph Merrick has a face warped by bone growth from his skull. His ears are small and shriveled. His right arm is very large, and his feet are club-like. Your highness, I must warn you that Mr. Merrick has some profound disfigurements. You must be prepared."

She was beginning to doubt herself. But when she saw the way her husband looked at her, waiting for an answer, hoping to have some lunch rather than look upon the famed Elephant Man, she said, "That I shall."

The Prince nodded unenthusiastically. "Whatever you want, dear. But let's do keep it short."

***

With the dedication ceremony behind them, the regal followers made their way down a narrow concrete stairwell. Dim light shone from wall sconces. The floor was clean, the walls whitewashed. The door to Joseph Carey Merrick's quarters had no lock or bars. It seemed, to the Princess, quite normal. But before anyone entered, Treves stopped the royal party short. The Princess looked across the faces of her attendants, as Treves now did. She saw a mixture of anxiety, fear, and anticipation.

"Mr. Merrick's rooms are somewhat confined. If it is acceptable to the royal couple, it might be best if only three or four enter."

The Prince chose the burliest of his attendants, and the four passed quietly through the door. The attendant entered first, carrying himself as if entering a battlefield.

The first thing Princess Alexandra noticed about Joseph Merrick's quarters was the lack of mirrors. Having lived under constant public scrutiny, she was used to checking her appearance wherever possible, primping until her fashion was absolutely perfect—and so the absence of any reflective surface easily stood out to her. Then, the smell hit her. It was the stuffy odor of sweat and something unhealthy. Against the far wall stood a bed unlike any she had ever seen. It was large and inclined at the head, with a brace on one side, perhaps to help the warped man enter and exit. At the end of the room, a doorway led to a basement-level courtyard. A small table and two chairs loitered in the next room, and it was from there that the famous creature emerged. The smell became so strong that she could taste its acidity at the back of her throat. She inhaled slowly through her nose, willing her stomach to settle.

She could not see him at first, but his shadow moved across the wall toward them, giving her an idea of Merrick's distorted visage. He peered from the dark room and, in a quiet baritone, said what sounded like "welcome."

The light gradually illuminated the face and figure of Joseph Merrick. Alexandra felt a winter chill enter the room, though no door or window had opened. His skin was a pasty white, like uncooked dough. His face had the appearance of flesh piled upon flesh with no rhyme or design. One eye nearly disappeared behind a lobe of skin, while the other seemed almost normal, with a neat eyebrow above. His right ear pointed toward the floor from his jawline. The left side of his face was less gruesome in some ways, with a well-placed ear and an even jaw. A crow's peak of dark hair had been combed back from his forehead. From the raw opening where his mouth should be, teeth protruded at odd angles. Perhaps most jarring was the suit coat, white pressed shirt, and necktie that the sad creature wore.

Treves cleared his throat. "Your Royal Highnesses, may I present Joseph Merrick."

Merrick bowed as best he could, teetering to one side. Treves and the Prince's guard both lunged for him at the same time, but Merrick recovered. A bizarre sound issued from his throat. The Princess thought he might be choking, but soon realized that he was laughing. Treves laughed also, and then the royal couple hesitantly joined in.

Merrick mumbled several other sounds that must have been words, gesturing toward the chairs in the other room.

"Yes," Treves said, "would you like to sit?"

Merrick pulled out a chair for the Princess, who took it gracefully. The Prince's bodyguard pulled out a chair for the Royal, but the Prince waved a hand toward Merrick. "Please, Mr. Merrick. I have been sitting too much this morning."

It was the first admirable act he had carried out in a long time, and the Princess smiled up at him.

Merrick let out a few more sounds. Treves interpreted, "Mr. Merrick is apologizing for the lack of seating."

A Little Ancient History

Merrick sat with a slow, practiced care born of pain and physical awkwardness.

"Nonsense," the Princess said. She hesitated, then reached across the table and touched his hand. "There is no need to make an apology. You are having unexpected company."

She pulled her hand back and noticed how Merrick winced, apparently unused to human touch. How sad. The flesh of his hand was soft to the touch, warm, and peppered with hairy warts and lumps. The meeting of their hands brought a remarkable change in Merrick. His face was so hard to read, but in his downturned eyes she saw a self-consciousness that the Princess did not expect.

Merrick seemed to growl, and to make a sort of chattering. She could just make out the words "too kind."

"You must forgive Joseph," Treves said. "He becomes emotional at times." Merrick nodded in agreement.

At that moment, the Princess noticed the shelves across the room, piled high with books. The dim light prevented her from seeing the titles, but she could see that the little library contained works of many shapes and sizes. Travel accounts rested atop thicker volumes, perhaps biographies or historical treatises.

The twisted man stood, his massive head bobbing to one side. He held up a finger and turned further into the room, bringing out a half-finished woven basket.

This time, Alexandra caught most of the words. "Would you believe that one such as me could ...like this?"

She looked at the basket. He handed it to her. Although the rim was as yet unfinished, with frayed straw sticking out in all directions, the sides of the piece were tightly woven into an intricate pattern. The bottom of the basket swirled in interlocking spirals folding into its center.

"I would indeed believe it, Mr. Merrick."

His lips tightened. He was trying to smile. He made several noises. Treves said, "It is his–"

"Second," the Princess interrupted. "Yes, I caught that." Looking into those forlorn, crooked eyes, she said, "It is an admirable work. I could not do anything like it."

Merrick took in a long breath, and the Princess strained to listen for his next statement.

"I like to learn new things. Don't you?"

She was taken aback by the frankness of his speech, but instead of anger, she felt kinship. "Yes, yes I do. I try to learn something new every day. I enjoy my horses, and ice-skating."

"And dancing," the Prince added. "Don't forget dancing, my dear."

"But this," she marveled. "This is complex. Elegant. Things like this take discipline and skill, perhaps beyond me."

Merrick shook his head. His shoulders moved with it, shaking the table and the teacups.

"Your highness, you could do it. I know you could. You seem clever. And really, it is a matter of..." The last word was lost upon her. She looked to Treves.

"Self-control, your highness."

A smile broke across her face. She began to laugh. "Self-control. Absolutely! I think all members of the royal family should learn this art. I must look into it."

"Yes," Merrick encouraged. He held up his warped hand. "If I can do it..." His voice faded.

Treves interjected. "Do you know the estate house of Lord and Lady Knightly in Northamptonshire? It is called Fawsley Hall."

"We have seen it," the Prince said.

Treves crossed to the small window. On the sill, he indicated a little house made of card stock. "Joseph built this from memory."

"It is a very good rendition, near as I can recall," the Princess said.

Merrick held up his hand and took care to form his words. "I am working on the Mainz Cathedral now. It will take some time longer."

"I can imagine," the Prince said.

"I have something for you," Alexandra said. She called to the Prince's attendant. "Please have Esmie fetch you a photograph."

"Yes, your highness."

In moments, he returned with a photograph of the Princess. She asked for a pen. Treves had a nurse bring a quill and ink at once. The Princess spoke aloud as she wrote.

"To Mr. Joseph Merrick, with thanks for your tea and hospitality. Signed, Princess Alexandra."

Normally, the Princess would hand an object like this to the attendant to pass to her subject. But she herself held the photograph out toward the man. Merrick's hand shook as he tentatively took the sheet from her hand.

"I will treasure–" Merrick could not finish the sentence. The Princess saw moisture in the deep folds of his cheeks.

"And we will treasure this memory," the Princess said, standing. "Thank you, Mr. Merrick."

Merrick stood unsteadily and bent at the waist as the royal party exited, the door closing on a very dark room made suddenly more luminous.

In the stairwell, Princess Alexandra turned to Prince Albert. "We really must learn to weave baskets."

"Like peasants?" he spat. She froze him with her stare.

"Like Mr. Merrick."

Part I
The Novel

Chapter 1
Gwen Baré

August 2229, The Present

They floated in Venus' golden skies, 30 klicks above an inferno of broken rock and simmering volcanoes, in a city suspended below a cluster of silvery gossamer balloons. Amber clouds rose into the gray-blue firmament like the pillars of a Greek temple, bending at their tops, great banners of mist flagging across the cloudscape. The air here was laced with toxic vapors that tinted the sky far greener than the skies of Earth. On the western horizon, lightning lanced through a scrim of sulfuric acid virga. McMurdo IV's gleaming hydrogen-filled envelopes loomed above the station, glistening in the hot Sun, undulating in the zephyrs.

But Gwen Baré didn't have time to take it all in. Not now: there was too much work to be done.

She cradled a small pile of sunflower seed husks in her palm, adding to it now and then with a *pfft* sound. She enjoyed the problem solving in her work. She reveled in the flow of energy issuing from bioreactors to the colony's infrastructure, the interface of biological systems with the simple mechanics of electrical ones.

"God, how I hate those alarms."

Gwen jumped, pivoting to see a bespectacled, stocky man in his mid-forties. He had a pate of rusty hair atop a honey-brown face, the countenance of the proud Maori.

"Aren't you the stealthy one?" she said, turning her attention back to a series of monitors. She tapped a screen, and the klaxons fell silent. She lay the seed husks on the counter beside a touchpad.

"Are you going outside?" the man asked. He was rubbing his temples.

"In a while. Gotta go downstairs first."

"Into the bowels."

"Those bowels keep your little outpost alive."

"So you keep telling me. How do you stand the stench?"

She waved off the comment. "You get used to that really fast. It's like crime scene investigators with dead bodies, I guess. I know you like your pretty solar panels, but we've gotten so good at microbial fuel cells. Besides, acid-soaked panels don't work so well, and my little microbes are sheltered from wind and rain and sun and insults. You should thank me." She said it playfully, tilting her head to the side.

"Oh, I do. I do."

Sometimes she could tell that Joshua Aotea was looking at her, really looking. Josh had that air now, and she knew what he was seeing. Gwen wore almost no makeup—her friends had told her on more than one occasion that she didn't need it, but she suspected they were being kind—and she never thought about her long, chestnut hair except to drag a brush through it absently in the morning. She always had a wayward bang or two dangling into her eyes, something an ex-boyfriend told her was "fetching." She was, in her own keen estimation, far less organized on the outside than in.

"Still, I see it more like the connection between zoo keepers and aromatic elephant crap," Josh added.

She smiled at him. "Now, the zoo analogy, that's spot on, considering our colleagues. Except the smell reminds me more of a bakery. Yeasty."

"A bakery never put out the kind of energy your little tubs do. How do they do that?"

Gwen assumed it was a relatively rhetorical question, but she sensed a teachable moment. "Think of it as a microbial fuel cell. We set up a biologically based electrochemical system. It uses bacteria to drive an electric current in the same way that bacterial interactions happen in nature. We're just borrowing technology that's been there for billions of years."

"Yes, yes," he held up his hand dismissively. "So you think you'll have our brownouts fixed by dinner?" Josh's New Zealand accent bled through, transforming the words into *brainouts* and *deenah*. Impatience edged his voice.

"You have some clandestine meeting at Gallagher's?"

Panic flickered across his face. She had come closer than she meant to.

"You might say that."

"You're the boss." She tried to sound cheerful and nonchalant. She dropped a few more empty seeds into the growing mound.

Josh scowled from beneath imposing brows. "Sometimes I wonder. And I wish you would refer to this place as a colony. It might be the smallest, but still."

"No worries. It's plenty impressive. I prefer it to McMurdo II. Although it might be nice if you had more than one store."

"Yeah, you have to go halfway around the planet to get any real shoe shopping done."

She was used to his chiding. It was half-hearted. He wasn't the chauvinist he pretended to be. She looked down at his formal footwear, completely inappropriate for an upper deck slicked with sulfuric rain. "I was thinking more along the lines of memory chips, but I'm sure shoes are tops on your list."

"Hey, a guy needs to dress up every once in a while. To remind people of his status in life. Right?"

"I stand reminded. You're not even wearing your cheery Joshua label. How will you remember who you are?"

He grumbled something inaudible and headed for the hatch.

"Have fun tonight," Gwen chirped after him.

"Ta."

McMurdo IV was a close community, and Gallagher's was one of only three watering holes. Mac Four was a place where secrets were hard to keep. And it was no secret that the power had been fluctuating for weeks. Despite her bravado, it had Gwen worried.

It was past her dinnertime, nearly past pajama time, when she discovered the infection. "Hey Calvin, have a look at this," she called to her night-shift apprentice.

"Always the mentor, Dr. G," the young man said, shuffling across the chamber. Peering over the rim of a large vat, he saw a river of lighter material wandering across the mealy surface of the microbe ranch. "Looks juicy."

Gwen nodded. "Yep. Anything that yellow is bad news. Got some kind of bad bugs in our colony. So what's next? You know the drill. Tell me why it happened, first." *Pfft* went a few more husks.

"Make you a deal, Dr. G. If I give you the right answer, you have to tell me about the sunflower seeds. You keep promising one day. Time to pay up."

"Deal. So tell me."

The student didn't hesitate. "No sunlight–"

"Yes, they like the dark. So?"

"So without the UV they're prone to mold or some kind a contagion."

"Very good. So what's the solution? What's the silver bullet?"

He frowned and looked at the ceiling, clearly stalling.

"We talked about it two nights ago," she encouraged.

He snapped his fingers. "A-4 antibiotics."

"Almost. A-4 is pretty harsh, and we'd lose twenty percent of our little good guys."

"Then that A-2J mix?"

"That's the stuff. Good juju."

"Did I earn a sunflower seed explanation?"

"I suppose," Gwen grumbled dramatically. "It turns out that I have a talent for microbes and a knack for fainting. I keel over at the drop of a hat. Runs in the family. Even my Uncle Jeffrey had it, and he was a rugby player. Keeping something in my mouth and on my stomach helps me stay vertical."

"Blood sugar thing?"

"Nope. They never found out what. But sunflower seeds are cheaper than drugs, so that's the route I take." She nodded toward the nearest vat. "Now, would you do the honors?"

"I got this," he said, lunging for a set of oversized syringes on a nearby counter.

She put a hand on his shoulder. "Carefully, Calvin. Carefully."

He took in a deep breath and nodded. "Got it, boss. Just like last time."

"Just like. Only without the spills. Neatness counts. Carry on, my good man."

***

It was late—too late for Gallagher's or Pepe's-Really-South-of-the-Border. But with any luck, the Dusty Rose would still be serving dinner, or at least substantial tapas. Gwen tied her hair into a pile behind her head, donned her suit, mask, and hood, and headed outside. At the end of a long day, she preferred walking on deck to taking the internal passageways. Even through the suit, she could feel the wind, hear it singing in the guy wires, watch the cloud tops boil beneath the station. The air was cool, just above room temperature. A row of wind turbines turned lazily along the edge of the platform. Supplementary banks of solar cells tracked the Sun, low in the sky. These days, new panels were so difficult and expensive to make; the return of solar would have to wait until more of Earth's infrastructure came back. For now, solar helped at peak times, but the bread and butter of the place was percolating away within her microbe vats.

The lightning storm to the west had settled down. Sunset would come in about a week of Earth days. The sky was tinting toward molten copper. *The Sun sinks slowly in the east,* she thought in a western movie narrator's voice. *Very slowly.*

Entering the airlock, she peeled her suit off, catching a faint whiff of battery acid-scented mist. She shook out her hair, hung her O_2 bottle in a locker, and proceeded into causeway 3. A few doors down, the cacophony of conversation and the lilt of canned music wafted into the corridor. The Dusty Rose was busier than usual. She stepped up to a dais. A hand-scrawled note read, SEAT YOURSELF. The sign seemed accusatory: *Alone again, my dear?* Gwen headed for a table in the back corner. A robo-waiter appeared, displaying the tapas du jour.

"I don't suppose you have any sunflower seeds?"

"Please request again."

"Number three and a coffee." She leaned back, closed her eyes, and rubbed the nape of her neck.

"Join you?"

Through one barely open eye, she saw Josh. She sat up. "What happened to your hot date at Gallagher's?"

When he shrugged, his neck nearly disappeared. "Got stood up."

Gwen flipped her hand at the empty chair. "Sit. This is a pretty small town. Social abandonment seems risky, in a group dynamic kind of way."

"She had her reasons. We had a little tiff, and I resorted to cheap sappiness."

"Yes, they say to always play to your strengths."

"Didn't work anyway." He turned to the robo-waiter. "Sirenium Stout, please." Turning back to Gwen, he said, "So how are your overgrown petri dishes getting along?"

"We understand the problem. Just have to wait a few days for it to cure."

He grinned. "Nice."

"I hope I'm not being overly optimistic. About our microbial power generators. I have to get to Mac Two before nightfall."

"The long night's coming up fast."

"Yep; you'll be out of your summer veggies here soon."

"At least we have some. Greenhouses on the greenhouse world," he mused. "Who knows, we might just let ourselves drift sunward when everyone gets tired of the dark. Of course, my researchers wouldn't like us leaving their ground sites."

"I heard Boston has had its second year of fresh asparagus. That's got to count for something."

Josh grimaced. "If you like asparagus. So, Mac Two. Into the sunlight. Taking the blimp?"

"It's easiest, I figure."

"Slow. If your little bugs keep you waiting for too long, just let me know and I'll check out one of the puddle jumpers."

"You'd do a thousand mile trip for little ol' me?"

"Those things are pretty fast when you floor them. And it's the least I can do for someone who's fixing our entire power supply."

"That's gallant of you, Josh."

"I'm not being completely altruistic. This place gets to closing in on ya. I could do with an outing."

The poor guy was hurting, she could tell. But she wouldn't pry about his embarrassing non-date.

Josh squinted, lifting a brow. "Say, didn't you go out to the Saturn system as a kid?"

"I went everywhere; my parents were nomadic. Went all over the place trying to help people put the interplanetary infrastructure back together, post Wentaway. It was a lot like being in a military family. If this is September, we must be on Ganymede kind of thing. So, yeah, we went out there to shore up things on Iapetus, and to see if Titan could be salvaged."

"Seems a bit overly ambitious now," Josh said.

"Well, that was before we lost the northern grid. That changed the balance. All of a sudden, Earth was a whole lot more needy. That's what we get for having a global village. As Earth's grid fell apart, everybody went home to mama, bringing their power plants with them. To 'pitch in' on recovering the global energy stuff." She used air quotes. "We know how that went."

"One domino falls. Earth had nothing to support the outer colonies, and Mars closed ranks with terra. Of course, they kept things going on a couple asteroids, but who wants to live on a dusty rock? They almost made it at Miranda, almost got to self-sufficiency. I was rooting for that, but I guess nobody had the infrastructure they needed out there."

"And won't, until Earth stabilizes again. Ceres is as far out as the tourists go these days. Independence in the outer system has always been one of those dreams that never quite came to pass. All those small outposts just couldn't make it on their own."

"Although Titan seems to be hanging in."

"Yeah, that's the story. I've heard things are pretty marginal there these days. Most of them gave up and moved to Mars."

"Seems like the preferred destination of most of those outer system refugees. And that uber engineer, Valentine, you knew her, didn't you?"

"Melciéna Valentine? My parents did. I barely remember her." It was just a little lie. She leaned toward him conspiratorially. "Why so curious? What's up?"

"The woman's resurfaced again. She must be five zillion years old by now. Here." Josh tapped his wristcom and streamed the news story to Gwen.

She frowned as she read the text.

Josh warmed to the subject. "It's been—what?—fifteen years or something? It says there." He jabbed a finger toward her screen.

Gwen scanned the story.

Josh peered over the rims of his mag-glasses, but the stream of data in the lenses still distracted him. He pulled them off and wagged them at the little monitor. "Weren't your parents close to her?"

She took her eyes from the report. "At one time. She was the one who really jump-started the Enceladus undersea colonies. Before the dark years, of course."

Josh wiggled his fingers as if playing an invisible keyboard. "Strange goings-on with gill people. Although some suggest she's nowhere near Saturn these days. I heard she was hanging out at the abandoned terminus on Miranda. In this thing–" he pointed at the screen–"they're guessing she's hiding out on Enceladus. I'm skeptical. Not much left on that snowball."

She shivered.

He noticed. "What?"

"There's something sinister out there, under all that ice. Things I don't want to see. Ever again."

"You were there? On Enceladus? With Melciéna Valentine?"

She nodded, but added nothing.

"Not close, you two?"

"As the article says, she kind of fades in and out. Back then, she was not nearly so mercurial." She smiled at something far-off. "She was a magician."

"Yeah, a genius at engineering. Heard she did everything from bioengineering to propulsion. Invented some new engine or something."

"She was a regular Renaissance woman. But I mean she was *literally* a magician. Hardly a genius with the sleight-of-hand, I guess. Disappearing coins, card tricks, that kind of thing. But Mom told me she was quite entertaining. They must have been quite close; Mom and Dad asked her to be our guardian if anything happened on their travels. Which, of course, it did. But later."

"How old were you when the accident happened?"

"There are no accidents. Just poor decisions. I was…older." She quieted. "But that was a universe ago. She veered off in other directions and so did I. I followed in my parents' footsteps, putting everybody's energy grids back together after Wentaway."

"You're still in that business, putting things back together, judging by the brightening lights here."

She shrugged off the compliment. "Small potatoes compared to Mom and Dad."

Josh lowered his voice. "Do you ever hear from the celebrated Melciéna Valentine?"

She jerked her chin up. The veins stood out on her neck. She spoke through clenched teeth. "Melciéna Valentine is a traitor to the human race."

Chapter 2
Interval 1: Grandma Baré's Diary

Autumn ca. AD 2206, The Past

It feels satisfying, somehow, thinking about the fatalities—about their ashes fluttering on the wind—in the same way that grinding your jaw gives you some kind of twisted relief from a toothache. But then the pain comes back and you realize it would have been healthier to just leave it all alone.

It isn't the furnace blast that took away the trees and birds and cars, and painted all the colors in gray. It isn't the tsunamis that barreled in and splashed away all those seaside resorts and little fishing villages and fancy embarcaderos. It isn't even the disappearance of the fruit and veggies and butter and cheese and fresh meat. What really hurts, what we've all gotten tired of so fast, is the dark. The cold. Ironic. In the west everything blackened and charred. Here, we just got ice and night. Not only did the gardens go, but so did the power. No place to plug in. No streetlights or desk lamps. All our light switches have those little stickies that say "Embrace Your Darkness." It's hard not to. Autumn comes earlier than it used to. Dead leaves in every gutter, singed by the AWOL ozone and piled against buildings, just like the papery corpses in the west. And the days are so dreary. Batteries are in short supply. Our neighborhood's resident crazy, Amelda, even resorted to candles. Where on Earth did she get candles? Why am I asking you, dear journal? The fact that I must write this diary in ink speaks volumes about the state of electronics and the skein net.

Then there are the plagues. Bird influenzas and cattle disease and the sickness that's run through the big cities on the eastern seaboard. One nice thing: I've stopped worrying so much about my bunions. Lots of animals have croaked in the greater Milwaukee area, livestock and wild deer and such, but that's just an inconvenience compared to the many loved ones, friends, cousins, parents and sons and daughters who perished in the great conflagrations or vaporized in that first terrible flash. Sometimes I even miss Uncle Ray. Never thought I'd say that in a million years.

They say things could bounce back fairly quickly. It's been a very long three years. I wonder what yardstick they're using.

An asteroid the size of Manhattan? I'd hate to see something bigger. This "moderate" rock visited a Biblical apocalypse upon us. Yea verily. Who knew that death from the skies could come in so many exotic forms? I suppose any self-aware Triceratops could have told us.

We travel through the house wearing blankets. We look like gorillas. Not very fashionable gorillas, at that. Will "fashionable" ever come back? Or sunsets? Or reading a good book by the bedside lamp to put myself to sleep? It feels like this will go on forever. We couldn't hear Wentaway from here (they could on the west coast, of course), but we felt it. We still feel its effects. Everybody cries at night. Why couldn't they stop it? The only thing I am certain of is that my two precious teens are going to grow up in a world very different from the one of my childhood.

But, as the good book says, "This, too, shall pass."

***

Wentaway was before Gwendolyn Baré's time. A lifetime before. But her parents had told her all about it, as most parents had told their children.

Why Wentaway?

Because, sweetie, that was when everything did. All those things that made life comfortable or easy or efficient. Electricity and light and heat and refrigeration and nice clothes and good food. It all went away, just like that.

People threw around lots of names for the event: the Bad Blam, Bigroar, the long winter. But *Wentaway* had a slightly whimsical edge to it, and in a world where whimsy was hard to come by, it became the popular term.

The impact's long-term effects killed millions and gutted the worldwide power grid. Failed crops and diving temperatures exacerbated the energy shortage. A permanent haze high up crippled the solar power centers and calmed the winds that powered so many of the turbines. When the Earth fell into disarray, so did many of the off-world settlements. Some, like the Martian network, stumbled only slightly and quickly regained their balance, after a fashion. Others—most notably the smaller outpost cities on Mercury, the Uranian satellites and Neptune's Nereid—packed up and rushed back to the home world, hoping to help out in any way they could. Still others dropped off the radar entirely, signaling a catastrophic collapse of their infrastructure. Nearly a quarter-century passed before the first transports began regular runs to the outer system again, first to Ganymede and then to Titan and Ariel.

Chapter 3
Carter Rhodes-September 2229

"I don't necessarily envy you the job you're taking on, but I do think you're the perfect candidate for it." The figure on the monitor let himself smile faintly, his stiff upper lip enthusing encouragement while retaining that all-important aloofness. His dense black hair lay close against his formidable head, pulled back in an oiled ponytail.

Carter Rhodes squared his shoulders. "Thank you, Premier Van Dijk. It's been an honor, and the progress we've made so far is simply fallout from me having a great team."

"A team that you put together, Colonel Rhodes. Rest assured, we understand that. But as you so succinctly put it in last week's report, we've got to step things up. Food and water inventories are bouncing back, as is the population, but those who control the flow of energy control the world. Our engineers in Japan and Korea don't know how long the East-Asian seaboard will last before there's a catastrophic something-or-other. They need 'energy supplement,'" he made air quotes, "to ease things, as I understand it."

"That's the case on a planet-wide scale. But I suppose you could make the case most strongly for that corridor."

"Strong case? The eastern corridor is a time bomb. It goes, and it takes a lot of other infrastructure with it. Cascades. Chains." The Premier shuddered.

Rhodes brushed his hand across his crew cut, a bristly plain of auburn hair. It was a nervous reflex that he wished he could break. "We'll just have to shore things up before that happens. I'm especially interested in some of the larger abandoned outposts; the Ariel mines, the undersea metropolis of Enceladus, the Pluto/Charon funicular infrastructure. Bound to be some salvageable generators of some kind. What we lack in power at home we more than make up for in space transport."

"Well, then, as my American wife would say, bring home the bacon, Colonel. I think the World Council chose wisely."

"Thank you, sir."

The man on the screen leaned forward. Several chins wobbled beneath his bulldog face. "And delegate, Rhodes. Don't do it all yourself, okay? It's a big solar system with lots of real estate to cover. You're in charge. Find the experts. Send out explorers. Have them report back. That may smack of traditional management technique to you, and part of why you're here is that you are *non*traditional. But some things do work, like delegating, okay? Tried and true. It will save precious time. That's not so hard, is it?"

"No sir. As long as I can find competent people out there."

"Yes, well, the Council has a short list of what they feel are brilliant candidates. I'm sure you'll do fine."

The Premier signed off.

Carter Rhodes dug the heels of his hands into his eyes and plowed his elbows into the desk. "Look at me," he mumbled. "I seem to have made it to middle management."

***

Adjutant Chevalier sat across the desk from Carter, his eyes as red as those of his superior. Blue light from the monitor tinged his face, casting it with the slightest hint of death and decay. He shook his head and stretched.

"That's a whole lotta names."

Carter grunted. "Any good ones?"

"A couple promising contenders. You?"

"Let me see what you got."

"Comin' at ya." Chevalier tapped his screen. In moments, five new names appeared on Carter's list.

"Five names," Carter marveled. "This is pitiful. There must be someone out there who's competent, at least."

"Number three looks okay, potentially."

Carter leaned toward his screen. "Potentially. Baré, Gwendolyn. Spent a childhood rocketing across the system. Mercury, Ariel, Enceladus." His eyes flickered in the screen light. "Studied…good grades…"

"Which no one really cares about after the fact."

"Shows she can finish what she's started, at least. And look at the experience. Organic power systems on three worlds. Mars, Europa, Venus. She knows that territory. Bioreactors have the kind of staying power we just might be able to take advantage of."

Chevalier's screen cast changing hues across his face. "Microbes to megawatts. And unlike half these people, she's still active in the energy industry."

"Let's see if you can track her down."

"Already got it, boss. She's still on Venus, hopping between the four McMurdo stations."

"I thought there were five."

"They lost one last year. Wind storm or something. Anywho, she's at McMurdo IV this week, about to set off for another station. The source I talked to was pretty sure she was heading for McMurdo II."

Carter smiled at his assistant. "Not bad for a Chief Warrant Officer, Armand. Let's have a little chat with her. She just may be our key to the Enceladus problem."

Chapter 4
Gwen

Josh's mouth opened and closed like a landed fish. Finally, he formed a few words. "Guess it doesn't matter that Melciéna Valentine won the Nobel Prize for genetics. Oh, and there was that little incident of the Braham Selfie Interplanetary Laurel for her Enceladan architecture." He shrugged dramatically. "To name a couple."

"She lost her way," Gwen said, deflating into a slouch. "She started out as this grand advocate for the struggling human race. Just like Mom and Dad, her position on all things was basically, 'Get the home planet in order and then worry about the rest.' Everybody had left the outer system by then, shut down Tartarus at Pluto, all the Europan outposts, the stuff at Uranus and the Triton waypoint. The military kept their small post on Iapetus, but otherwise things were getting pretty quiet out there. Enceladus held on for awhile. Then, once they abandoned Enceladus and Iapetus, something broke inside her. She changed. Started playing God-in-a-petri-dish. Dropped off the grid for awhile. And then she got in touch with me – this was after Mom and Dad were gone – and started in on how I needed to be more open-minded about resources. How Earth needed to start sharing again. It was as if she hadn't seen any news feeds in a decade. The eastern corridor on Earth is still reeling, on the verge of a new Dark Age, and she's talking about sharing? With whom? It was completely counterproductive. Sure, she was theoretically our guardian. But I thought the entire communiqué was pretty cheeky. We—the Earth—really could have used an engineer of her talents."

Her wristcomm flagged an incoming message. She blinked at the microimage while Josh continued. "Maybe she was just shifting her focus a little. Seems like we could heal the home world while we're sharing with the other settlements."

"Spoken like a true administrator of the Venusian outposts."

Josh offered a wan smile. "I suppose. But life is not so black and white."

She didn't take her eyes from her wristcomm. "That's what my sister always says," she mumbled. "I'm sure Melciéna Valentine would agree." Josh waited. Finally, Gwen said, "Do you ever get the idea the universe is trying to tell you something?"

"Often. It's usually something I don't want to hear."

Gwen looked up at him. "I've just been invited to a meeting with Carter Rhodes."

"A face to face?"

"With coffee and crumpets and the whole nine yards, I'm sure."

"He's a VIP in anybody's book." Josh took in her frown. "And here you sit grumbling about it. Why?"

"I hate going back there. It's so sad."

"The mountain won't come to Mohammed. Besides, all-things-Earth are getting better all the time. Remember what you said about asparagus."

She glanced through the window. The clouds were tinting toward purple as the days-long sunset unfolded. Unchecked, the station would be blown around the planet in less than a week at this altitude. By sinking and rising to catch the currents, the buoyant city could remain in place or travel the world.

Gwen closed her eyes. "There's so much destruction left. You can just feel how things used to be. Sometimes I look at the ruins and the garbage piles and I just think they've given up."

"Certainly not everyone has. And this Carter Rhodes seems to be pretty enthusiastic. He's on a voyage for voltage. Treasure-hunt for terawatts. Quest for current."

"Your management degree was wasted. Where was the English department when you needed them?"

"Can you make it to Mac II before you leave?"

"I'll have to move my schedule up."

"If you're sure we're okay down in those bacteria bins of yours, I'll take you."

"Thanks, Josh. You're the BB."

He leaned forward. "Big Bang or not, are you sure we're going to be okay? Despite their aroma, I love what those little beasts do for us."

"My able assistant can let me know if we're not in the clear; I can make a mad dash back before I leave for Earth if I have to. But by tomorrow things should be up to speed, and you can give me a ride with no worries. Good?"

"Good," he said with thinly veiled skepticism. "I don't get out enough, and I like to see what the neighbors are up to."

***

"Don't you wish this thing was a convertible?" Josh hollered into his microphone. "Just put the top down and feel the breeze?"

"I'll bet the last time you did that was in an oxygen/nitrogen environment. Your experience here would be less…entertaining, I'm sure."

He bared his teeth in a joyful grin. The bow shock spread the mist ahead of their craft like Moses parting the sea. Behind, Josh was undoubtedly leaving a series of sonic booms. He pointed down. "Looks like thin overcast today. Should we have a peek?"

"Sure," Gwen said. "Always love the chance to see Dante's inferno down there."

Josh banked left and dropped his noisy little vessel through a scattering of high clouds. The mist below was light today, and the sunlight made its way well down into the planetary soup. Josh continued his descent. As the illumination dimmed, the haze below began to shift from yellow to gray, and finally to a rich umber as the surface appeared and the sky vanished. The horizon melted into the mist as the panorama under them cleared into a tortured blanket of basalt. Pillars and swirls of ancient lava spread before them. A field of magma had cooled into a vast taffy-pull of twisted, burnt stone. Black sand dunes banked against rust-colored, rocky berms. Tarry lava washed up against low foothills like a seascape frozen in time. A dramatic fault broke the undulating plains, heading off toward a distant, low butte, barely visible in the sulfuric acid haze.

Josh pointed. "Nice pancake dome. We're coming up on Eistla Regio. Should be a lot of them."

Gwen had always liked the pancake-shaped volcanoes, with their flat tops and steep slopes. The mountains were only one type in the menagerie of volcanoes that Venus hosted: ticks, anemones, shield volcanoes, majestic cones.

Josh pulled the nose of his little hotrod up toward the luminescent haze above. As they pierced the cloud layer, McMurdo II loomed above, its silvery envelopes glistening in the sunlight like a hot air balloon convention. Most of the high-flying village rested atop a series of platforms, each topped with glass domes and columnar habitats resting on their sides, the balloons rising from the decks like a cluster of colossal mushrooms. Windows beneath the top deck betrayed the presence of other levels farther down. Wind turbines cartwheeled along the edges of the top decks. Somewhere within the lower stories, Gwen's objective simmered in vats, sending its power coursing through the station.

Josh cleared the rim of the landing pad, pirouetted the craft around by ninety degrees, and let it settle on the deck. A conveyor slowly pulled the vehicle into the hangar. Overhead, a door rattled down, surrounding them with a transparent half-dome. The air cleared, green lights flashed, and Josh slid the canopy open.

Gwen yanked her mask off and then pulled her goggles down, letting them drop to her collar. "Buy you a cup of coffee? The bar here does an excellent brew."

Josh gazed through the dome at the wind turbines, the guy wires, the stacked landing pads like the cards of a leaning deck. "You're on."

She sensed the unique ambience of Mac II instantly. Where McMurdo IV gyrated with the excitement of discovery, thanks to its fleet of observatory drones and army of cheerful visiting researchers, this station simmered in quiet, constrained solemnity. Maybe it was the smaller population or the morose Station Commander, known for his iron-fist supervision. Maybe it was just the subject matter: this was an atmospheric station like Mac IV, but the target here was the depths of Venus' carbon dioxide ocean of air, down in the "dungeon" near the bottom of the Venusian sky.

They sat at a table by a wide portal, lattes in hand. Outside, a bathysphere lowered past them on a cable, destined for the high-pressure world below. Across the flank of the sphere spread bold lettering: "Broadwalter and Associates."

Gwen watched the sphere disappear in the fog below. "Broadwalter?"

"Reginald Broadwalter," Josh said into his nutmegged coffee foam. "Rich guy who hangs out around Miranda somewhere. He's quite the patron of the sciences. Gives lots of his money to our Mac constellation here, like that pretty little metal bubble those poor sots are taking down toward the surface. It's a long way down. I don't think I'd go. Lightning and acid and dark and pressure." He shivered.

"We just did."

"It was entertaining, wasn't it?"

"Thanks for the ride, Josh. Made my life a lot easier. I should probably get going."

He tipped his head, looked at her sideways. "You know, Gwen, you are a driven woman."

"Thank you," she said, folding her napkin.

He winced. "It wasn't meant as a compliment. I feel like you're being dogged to produce, as if someone's peering over your shoulder all the time. It's like at the end of the day you have this balance sheet where you want your output to exceed your input, and I have no idea what it would mean if you didn't make it, because I'll bet it's never happened before." She was silent. "Maybe it's because of your business: energy in, energy out. Maybe it's because of your parents."

"I had great parents!" she snapped.

"I'm sure you did. That's not what I meant. In fact, that might even be part of the problem." He tried to correct himself. "The issue– This isn't coming out right." He folded his hands on the table.

Gwen couldn't help but feel sorry for the guy. Josh always seemed to be saying the right thing the wrong way. Maybe that was what went wrong with his recent date.

He began again. "Sophia and Hugh Baré were pretty famous, running helter skelter across the system, plugging holes in the cosmic dike and shoring up energy here and there and saving the Earth. It's a tough act to follow." When Gwen said nothing, he added, "What you do is important enough. You don't have to be like them."

Gwen took in a long, slow breath, puffed up her cheeks, and let it out. The slight exertion made her dizzy. She closed her eyes for a moment, then said, "Guess it's my takeaway from childhood. We watched them for years as they played cavalry, galloping in where life was a disaster, magically repairing and fixing and revamping. Their work was important on an entirely different level. It meant something, large-scale."

Josh kept a gentle tone. "As opposed to yours? Is that what you're saying? Because if it is, you're forgetting how you just saved the power supply of the entire McMurdo IV colony. *My* colony. And you're about to do something like it here. You can't judge your worth by how many people you've saved from some disaster or crisis. It's not about notches on your belt or numbers of noses. You do great work, Gwen. Just let it be what it is."

She smiled weakly. "I appreciate that, Josh. You're a good man. The younger uncle I never had." She stood. "And a good chauffeur. See you when I get back."

"I'm assuming you'll bell me up on your return to our fair city? Be careful out there. It's a big universe."

"And miles to go before I sleep." She leaned over and kissed him on the forehead. "Thanks for the chat."

As she left, she heard him mumble, "Fat lot of good it did."

She made her way toward her quarters. Was he right? She could be passionate about the world's power supply, or about the microbial fuel cells at the heart of

McMurdo IV, but how long had it been since Gwen had lost sleep over another person? Could she even do it? Why could she not care, *really* care, about others? Was there something missing in her psyche, some part broken ever since she lost her father?

It always took so much energy to live an examined life. Perhaps sleep would wash away the complexities of her universe, her history. She went to bed.

***

As way stations went, the orbital hub a thousand miles above Venus was humble. It offered three docking ports, but only one was rated for the big cruise ships. The place was utilitarian: short on décor and long on bolts and cables. The floors were, for the most part, metal grates, and the walls huddled close together. Still, it got the job done, with ships trundling back and forth to Earth on a dependable schedule. Gwen's stopover there was a short one; her ship was waiting when she arrived.

The three-week trip from the Venus orbital hub to South America's posh Spaceport Sao gave Gwen plenty of time to think. Time for the ghosts to visit and the doubts to creep in. What did Colonel Rhodes really have in mind? What role could she play? She'd been out of the mainstream, off on the hothouse world, for a very long time. If she was going to be useful, she should probably brush up on "current" events.

Current. She smiled at her little science joke as she culled the news reports. Her screen lit the tableware next to her, its soft blue light reflecting back from her fork and spoon. Despite her misgivings, excitement knotted her stomach. This was a chance to make her mark in a big way. Josh had been wrong; keeping the infrastructure of Venus together was a minor aside compared to the work still needed on the genesis world. Rather than servicing a quartet of science stations in the clouds of Venus, she could make a difference in thousands of lives, maybe millions. Like her parents did.

The news feeds brought back the old melancholy. It would be decades – or even centuries – before the antebellum mansions returned to the southern portions of North America, or the Victorian architecture rose again from the northeast. The Golden Gate finally stretched across San Francisco bay again, but Oakland was still in shambles. The rock that fried the entire northern seaboard of Australia had also singed California and Baja with remarkable efficiency. Some of the haciendas were back, and there was trendy rebuilding in the suburbs of Laguna and Newport. They'd even resurrected the casinos on Catalina Island, once the waters receded. The submerged islands and raised coastlines appeared again, slowly, as the southern polar ices began to refreeze. But nobody had steady power, even now.

By the end of the voyage, Gwen felt as though she was nearly up to speed on all things terrestrial, at least where the energy budget was concerned. The story was dismal, but that only spurred her on. Driven, Josh had called her. Maybe so.

The ship docked to Earth's geostationary orbital hub and, grabbing her gear, she transferred onto the shuttle to Sao Paulo, NeoBrazil. It was late May, and the snows had not yet come to the southern lands. But they would, and with fury.

Chapter 5
Rhodes

It was reunion central as Gwen stepped from the space cruiser. People laughed, hugged, and slapped each other's shoulders. No such celebration awaited her: this was business.

A slender, muscular man stood beyond the fray by the ticket counter across from the airlock access. His flat top was graying at the sides. He wore a shirt tight enough to show his well-built physique, but loose enough around the collar to scream "casual." Carter Rhodes must not run that tight of a ship, she mused. For the adjutant of a military man, this one was a bit nonchalant. The man straightened and stepped across to meet her.

"Gwen Baré," she announced, shoving her hand toward him.

"Dr. Baré, we appreciate you coming. You took a long trip on short notice."

"Hey, when someone like Colonel Rhodes calls, I go."

"I'm not sure how to feel about that. You make me sound like a tyrant."

"You?" In that moment, Gwen realized that all Rhodes' communiqués had been text—non-visual. She'd had no warning of what the man looked like in person. "I'm so sorry, Colonel. I thought you were…"

A smile slowly broke across his face. "An assistant? Underling? Yes, life is full of surprises, I suppose. The apology is mine. I had taken the afternoon off and decided to fetch you myself."

"Hardly the work of a tyrant," she said.

"Let's hope not. Shall we?"

The two took the bullet train on a six-minute trip across the Sao Paulo/Brasilia metroplex to the Association of Worlds Councils. The AWC campus included several glistening glass and steel edifices reaching into the clouds, along with manicured parks and canals. The place showed none of the shabbiness Gwen remembered of Earth. In fact, it sparkled. The train stopped directly adjacent the elevator towers. Next to the cab's call button, someone had plastered the famous red-and-purple sticker that read, *Embrace Your Darkness*. As they stepped into a lift, Rhodes called out, "Six thirty-four."

Six thirty-four the elevator voice repeated. The cab accelerated gradually, but once they were up to speed, Gwen could tell they were traveling at a good clip. Their rapid passage through the cloud deck reminded her of the ascent to the Venus orbital hub. The lift chimed and announced the destination. Rhodes followed Gwen out the doors and down a well-lit corridor with plush carpeting and mirrored walls. Several people saluted as they passed. Every power switch or monitor had a little red-and-purple sticker, reminders to conserve energy. Rhodes stopped beside a large door that irised open to let them through. Inside, a traditional double door of pseudo-wood opened into the office.

They stepped into the room, a chamber dimly lit save for the panoramic windows across the back wall. Gwen could see the horizon out there in the haze, with low-lying green mountains and the sun glistening on the distant sea. "How high up are we?"

"Mile and a quarter," Rhodes said. "'Bout two klicks."

Gwen plopped down in a posh chair next to a huge faux mahogany desk. She saw spots glimmering in her vision, but they faded rapidly.

"You okay with heights?"

"Sure, fine," she said.

"It's just that you looked a little teetery for a second there. Water?"

"I'm good, thanks." She popped a couple of sunflower seeds and felt more stable.

Rhodes took a seat on something that more resembled a royal throne, surrounded by embedded monitors on the desk's far side. Behind him spread shelves with models of ships, ranging from ancient water vessels to modern interplanetary frigates. At the far left was a colorful miniature of a Minoan Admiral's ship, with its great sail, blue and orange stripes, and fifty oarsmen. To the far right stood an interplanetary yacht, outfitted with sleek fins and a streamlined cockpit.

Upon the corner of the desk lay a glossy photo album. Gwen gingerly brushed her hand across the cover. "Funny how the old things come back."

"I love photo albums. They take no juice, except for the ambient light that's already around. Guess that's why libraries are coming back. A generation ago, nobody knew what a conventional library was. Instead of a keyboard, a great, wooden double door with brass handles and frosted glass. Parades of multicolored volumes along polished shelves, their covers brought to life beneath a wash of spotlights." He smiled, almost sheepishly. "My nephew actually works in one. Be my guest, please." He gestured toward the thick tome.

She opened the book carefully. Traditional prints and holophotos plastered the pages with three dimensional figures: children playing in a sprinkler, lovers holding hands, a tourist scene at New Venice on Europa. Days gone by. Scraps of polychrome fragments decorated a few borders. She brushed her finger across one. It held the unmistakable feel of pulp…real paper. Family. Memories trapped within paper.

Her family had no such book.

"Beautiful," she said, closing the album softly.

Carter folded his hands on the table and studied her for a moment.

"Dr. Baré, we're here to talk about microbes and moons."

So much for small talk. "I can certainly talk to you about microbes."

"Yes, your record shows you to be quite capable on that count." He leaned slightly toward his central monitor. "Miss Talus, could you please send over the secure file marked A415? It's under clearance level two."

"Here it is," a voice said.

Rhodes typed in a security code and a display came up on the screens. "Do you recognize this curve?"

"Energy output," Gwen said immediately. "Not nuclear. Certainly not solar or wind. I would guess it's a product of bioreactors, judging by their curve, and also by the fact that you brought me all the way out here."

"Yes. This particular one defines the energy output of the microbial power plants on Enceladus just before Wentaway."

Enceladus. The mystery world. She fought another shiver and steadied her voice. "It's a very healthy output, I would say. As robust as a good day on Venus or Earth."

"And what are the chances that it's still got a healthy output, after all these years?"

"I'd have to look at the automation level of the facility, but I know they kept things on Enceladus going for a decade or more."

"Yes, we had an intelligence report a few years ago. One of our sources suggested that Enceladus engineers had come up with some kind of super-fuel cells."

"I heard those rumors, too. If the hardware was fairly contemporary, most of the vats should still be intact. At least dormant."

"I had hoped that was the case." He paused, hit a button, and watched all the monitors fold into the tabletop. He stood and walked to his window, grasping his hands at the small of his back. Cloud cover obscured the landscape seven thousand feet below. With a bit more yellow and a hint of sulfuric acid in the air, Gwen would have felt right at home.

Rhodes kept his gaze toward something outside, something distant. "Dr. Baré, the Earth is in trouble. More trouble than we have let on to our dear public. You see, we've built ourselves a straw house. In place of straw, we've given ourselves the energy of a thousand bioreactors and a million solar panels and windfarms. But it's the bioreactors that we came to rely on, especially with all that asteroidal dust dimming things."

"And killing the wind," Gwen added.

"Well, it really shifted more than killed it. The currents are all messed up, along with the ocean currents—they say those two are chained together—and the wind power plants are all in the wrong places. At least for now. Of course, it didn't help that the rock landed on some of the largest power generation centers on the globe."

"Yes," she nodded. "All our eggs in a couple baskets."

"And all that lack of a diverse energy system—all those eggs—are cracking under the stress. We've seen rolling power-outs before, but we seem to be gearing up for a failure on a much larger scale. Something big is coming, something that could blow that straw house down, and I plan on stopping it."

"Sounds ominous," she said.

He crossed to his chair, and stood beside it. "Believe me, it could be. I'm not being melodramatic. The Eastern grid is hanging on by its teeth, and it's not just about infrastructure. To get it back up and running, it needs a supplementary infusion of energy, and it's energy that we just can't generate right now. We've dug up all the nuclear fuel that was disposed pre-Wentaway, but even that has to be retrofitted and fusion-rated, and most of it is trash by now. And as for bioreactors, you know how long it takes to establish a good array of microbial fuel cells. All that culturing and seasoning and nurturing that you experts do. To make the grid self-sufficient again, to stabilize the planet's overall power network, we have to get some juice from somewhere else. We have a handful of reactors—biological as well as nuclear fusion—on their way from the outer system. But we've only located a few, not enough to do anything substantial."

"And what happens if you can't get more?"

"Have you ever watched a house of cards when you put on one too many?"

"Embracing the darkness," Gwen mused.

A grim silence stretched between them. Finally, the Colonel added, "And then there's Enceladus."

Gwen frowned. "Enceladus." Rhodes nodded, not smiling. "Colonel, I'm a bit confused. I thought that leaky little moon was deserted. Has someone gone back?"

"Nope. No reason to. The entire population 'abandoned ship' decades ago. Ran home to mama Earth. There was no reason to resupply the undersea cities because nobody was left."

"And the Naiads?"

"Left behind in the last-minute panic at Enceladus. Not by design; there just weren't enough ships to get everything out. Some of the inhabitants planned on going back, taking care of the surviving ones as best they could, but they had no idea how tight the energy crisis would make everything."

"Including travel back to the outer system."

Rhodes nodded. "Oh, I imagine some of the hybrids survived for awhile, but they really didn't have the wherewithal to take care of themselves. Sad, really. They were interesting. Something out of Greek myth; those supplementary gills, those webbed hands and the spidery fingers–"

But they weren't real. Not in a natural way. As a child, Gwen thought of them as not-very-pretty mermaids. It would have been different if they were a natural occurrence, some primordial phantom that crawled out of the ooze. But as she got older—and a bit less naïve—she saw them for what they really were: a thing warped from the natural, twisted to a darker purpose, an abomination.

"I saw some. When I was young," she murmured.

He sat down, his eyes at her level again. "Me, too. They were quite dexterous. As long as they were given some guidance, of course. I actually saw several when I was on assignment to Triton. Stopover at Atlantis West. Rather remarkable."

"The city?"

"Everything. The Naiads had large heads that sort of bobbed when they spoke, and because of the gill structures under their jaws, they were difficult to understand. They breathed air normally, but their nostrils had this strange slit shape—they could close them underwater—and their chests were oversized so they could keep extra air inside whenever they transitioned from lungs to gills. Gave them a sort of rasp, a gurgle, in their voices. Looking at them, it was hard to see the human part. The slits in their noses and those recessed ears…those genetic engineers didn't have their minds on the aesthetic, certainly."

Her memory chased the cool green of the skin, the scaly texture, the overlain slime scattering the light. She fought down a shudder, and mumbled to herself, "Why didn't they just use construction bots?"

Rhodes heard. "They used them, too. But the hybrids could think on their little webbed feet better than any AI program available; problem solve on the fly. They were flexible and coordinated underwater, far more than the most fluid of our AUVs."

When Gwen frowned, he added, "Automated Underwater Vehicles. And believe it or not, sub-human clones are cheaper to make than a top-tier AI construction bot. Even with the food bills. Imagine you bash your AUV into a rock or some girder. You have to repair cables and wiring, replace electronics, fill in and sand the cracks. With a hybrid, you slap a bandage on it, give it a day to rest, and you're back to work. They were also supposed to solve some manufacturing hurdles by being self-replicating, but that didn't work out. Something about genetic sterility in a cloning environment. Let me tell you, for brutes with the intellect of a three-year-old, they did beautiful work."

Gwen found herself growing uncomfortable. She changed gears. "When I was a kid, I saw the primary seafloor structure, the big thing that looked like a butterfly."

Rhodes nodded enthusiastically. "Yes, yes, impressive engineering there. The walkways along the edge of those wings had floodlights so you could watch the sea around the place. But I was more impressed with the interior. Did you make it in there? There were trams and soaring structures and moving walkways. Everything glistened. It was engineered by some of the best, and built by simians. Ironic."

Simians. Certainly not a generous view of these cloned servants. In quite the contrast, she could remember her honorary Aunt Melciéna describing the Naiads as masterpieces. Rhodes didn't know the half of it. He didn't have the personal insight, the advantage of Gwen's experience, of her family history. Innocence could be found in many places, but not, surely, among the Naiads. Her discomfort still growing, she tried again to change the subject. "So…you're in search of bioreactors in a sunken colony?"

"Two colonies: Atlantis West and Thera." She could see fire in his eyes. "Since the twin settlements made extensive use of bio-energy, some transportable sources may still be active. In fact, I'd bet on it. We sent a fleet of satellites out to measure energy flow at various abandoned sites, including Enceladus. Can't tell anything in the south because of all the geyser activity. It interferes with the sensors. But further north, in the deep-sea regions where the settlements got plopped down, there seems to be a positive energy flow, and it's not the magnetic field of the oceans. No, there's still something putting out power there, and my guess is it's something we badly need. Some historians suggest that the inhabitants took all the nuclear with them back in oh-eight, although there wasn't much to take. On a dark ocean floor, that leaves only one viable option."

"I see where you're going with this." Gwen said. "The two settlements had the most advanced bioreactors anyone had developed up to that point, right? I've seen the old schematics. Our newest systems aren't much more advanced. If they really built what I saw on the blueprints, then yes, it would be worth looking into."

"You've been doing a little homework. And transportation?" the Colonel asked, leaning forward.

"Expensive. A little tricky, keeping the microbial colonies viable all the way back to Earth. They probably wouldn't make it on one of the cycling ships. Too slow. A transport is more like it. Expensive, but worth it."

"Of course. How can one put a price on the salvation of humanity?"

Gwen heard her father's voice within Rhodes'. "You can't. Human life is the ultimate treasure, absolutely."

"Over all else?" *Was he quizzing her?*

"I can't imagine what else there would be."

"Yes. We've got work to do if we're going to reestablish our standard of living before Wentaway."

"People are still dying from the power failures," Gwen pointed out.

Rhodes became pensive. "Yes, yes. We've all lost loved ones. All of us."

She could tell the comment was not hypothetical, not general. It had the ring of fresh, raw ache to it. Should she pursue it? Was that too forward? Perhaps so, for now. So she simply said, "If things get worse, more will. The medical facilities, the centers of learning, simple things like water treatment, all teetering on the brink."

"And if we can't shore up that grid before it fails, we'll have another global disaster on our hands. We've already had outbreaks of Cholera and Typhoid and some strange stuff from the rampant fungi. If we lose our infrastructure, we'll be back to campfires in caves."

Gwen watched him. They were on the same page. "Most people don't understand that power is the key to it all. The human race fought long and hard to get where it did, and we can fight our way back again," she said. She tried to read the Colonel's face. What did she see there? Admiration? Elation? It was difficult to tell, because he was up and moving again.

"So Dr. Baré, to strategy. We've got you booked on a flight from the Venus station to L1, just a million miles from here, give or take. You board a fast transport from L1 to Phobos. We'll get you on a cycler from there direct to the civilian hub at Iapetus. We have some resource limitations in that area just now, so a cycler is the best we can do."

"A bit slow," Glen mused.

"It is, but it'll have to do for now. One of our military shuttles will ferry you over into Enceladus orbit, but we'll have a big transport standing by to return you directly to Earth with your cargo, fast. My suggestion for a first stop is Thera. It's the smaller of the two major settlements, but we're hoping there is still one of the supra-nuclear fusion jobs there, probably offline. They called it Big Bertha. And my sources are almost positive that the two smaller fusion plants are still operating, low-level. These are also the advanced supras, so well worth going for. Not common knowledge. You'll naturally pick up any active microbial fuel cells there, too. That's the pay dirt. After Thera, you can stock up on what's left at Atlantis West."

"The bigger settlement, right?"

"Correct. Is that where your father–" The question died in the air. Gwen nodded in a nonchalant manner that she didn't feel.

"Well," he said. Standing against the panoramic windows across the room, he grimaced.

"Something else?" she asked.

He waited for a moment, and then stepped over to tap a button on his desk monitor. "Simpson, can you come in, please?" He looked up at Gwen. "You should hear it from an expert. You see, I am successful only because I surround myself with excellent advisors. That's why you're here for the Enceladus project."

A woman in full uniform entered and stood at attention. "Sir."

"At ease, Commander." The military officer spread her feet slightly and shoved her hands behind her back. She didn't look at ease to Gwen.

Rhodes continued. "Dr. Gwen Baré, this is Lieutenant Commander Darlene Simpson. Her expertise is interplanetary communications. Commander, can you share with Dr. Baré our challenges at Enceladus?"

"Yes sir. Certainly, sir." She stopped looking at the ceiling and locked eyes with Gwen. Her disciplined military face began to soften with enthusiasm. "It's quite mysterious, really. Saturn is unlike Jupiter in that its magnetosphere is fairly tame. While Jupiter is busy frying most of its inner moons and the majority of our low-orbit equipment, Saturn's radiation environment is calm. All except for the neighborhood surrounding Enceladus," she said, her disciplined military face softening with enthusiasm. "Communication in the environs of Enceladus is frequently disrupted by an as-yet unidentified local source somewhere within the moon itself. It seems to have nothing to do with the southern cryovolcanism, and we've tried to tie it to seafloor volcanics, but it just doesn't fit. It's sporadic and unpredictable to this point. We have the computer modelers working on patterns, but nothing explains it so far."

"They think it's coming from the core?"

The Commander grinned. "That's the thing: it seems to be coming from within the ocean, very amorphous. Not like a lighthouse beam. More like a blaring foghorn. It's all over the place."

Gwen needed sunflower seeds. Now. Surreptitiously, she pulled a few out of her pocket. "And this just started happening?"

"There were hints of it, even back in some of the early robotic exploration. But nobody really noticed until people started settling the Saturn system."

Rhodes took up the narrative. "Yes, so we have this mystery that keeps getting in the way of recon sats and any piloted forays out there. I suspect it may have something to do with the power output of those bioreactors. Dr. Baré, what do you think?"

His suggestion contradicted all of her experience. She had worked with many models of bioreactors, large and small, on Venus, Mars, and the Earth's Moon, and never had a hint of radio emissions from her microbial colonies. That's what she was thinking as she chewed her pips, but what she said instead was, "It is possible, sir."

"Thank you, Commander. You're dismissed."

Lieutenant Commander Simpson saluted, turned crisply, and departed. Rhodes frowned. "This will not do."

"What is it?" Gwen asked. It had all seemed to be going so well. What had she done now? Perhaps it was the informality of her sunflower seeds. They'd gotten her into trouble before.

"This 'Sir' bit. Please call me Carter."

"Yes, s—Carter. You can call me Gwen. Or 'your highness.'"

They both laughed, he easily, she tentatively. She did not expect to be on a first-name basis so soon, if at all. But this Carter Rhodes was a manager of a different kind. More than that, he shared her sympathies and priorities. Improving the human condition took precedence. No matter what. That was all well and good, but she knew it would probably lead to a very long, very cold trip to a very wet world embedded within the E Ring of Saturn.

Rhodes closed his eyes against the sunlight streaming through the window. "Just a little more power for the mix, and then we can do things. Not just survive, but live."

Clearly, Rhodes' thoughts were far away again, on something not defined, something enigmatic. Gwen had to ask, "And what would you do with that extra power?"

His eyes flew open, as if she had surprised him from a sleep. "Me? Personally? I do have dreams, but they're of no consequence."

"Sometimes our dreams are the very thing that enables us to forge on, to win our battles."

"No doubt," he said.

She looked at him, her smile encouraging him to finish.

He looked down at the floor and spoke like a self-conscious schoolboy. "You know, I've always loved the mountain towns in the Rockies. Particularly those towns along the summit corridor…Breckenridge and Aspen and Vail Village and Hickenlooper Hamlet. Now that things have settled down a bit, the clean snow has returned to that high country. Snow brings back childhood memories for me. Shooshing."

"Shooshing?"

"Skiing. Ever done it?"

"They have something like it on the Europan ridges. When I was a teenager, we sledded down the parallels of Thinea Linea. You have to use heated runners; otherwise there's no slip to a surface in a cold vacuum. Works great. But I never tried skiing. Is it hard?"

"You have to get the hang of it. And I did. You slalom back and forth like a fish, or like a bird, with the wind in your face and the trees rushing by. And the skis make a sort of shoosh against the snow."

"Shooshing," she said.

"Shoosing," he smiled back. "I went back to Breckenridge late last year. It's a ghost town with grand possibilities. When we get power to the high country again, I'd love to bring the winter sports back to Breckenridge. I'd love to see those slopes filled with people again, racing and jumping moguls, enjoying the snowy slopes."

"You see?" Gwen said. "It's a nice dream. A good reason to win our battle."

He stood. "Yes, well, speaking of dreams, we need to get you checked into the hotel. Adjutant Chevalier will get you all squared away. I've got a three o'clock or I would take you myself."

She waved him off. "No worries. Thanks for the chat—the briefing. I'm looking forward to playing my part."

"And I'm looking forward to us working together." He said it with gravity, but with no sense of sexuality. She had learned to recognize the misplaced romantic angle the hard way, and in many contexts. *Good*, she thought. *This is business, after all.*

"Details tomorrow," he said. "Get some sleep."

She did sleep. It was a good, deep sleep, the kind that brings dreams. But her dreams were not sweet. She was in darkness, with the smell of dank seaweed and the sounds of dripping water, the screams of her sister. A giant eye watched over her and, just barely discernible, a green, scaly hulk moved toward her.

Chapter 6
Interval 2: A Fairy Tale

Once upon a time, there lived an old mother pig. She had three little pigs, and not enough food to go around. So when they were old enough, she sent her three little pigs into the wide world to find their fortunes.

The first little pig was very lazy. He wanted to finish his house quickly, so he built it out of straw. Then he sat back, relaxed, and sang himself a little song.

The second little pig was not so lazy, but he was still too lazy to build a good house. He built his house out of wooden twigs. Then he sat back, relaxed, and sang himself a little song.

The third little pig was a hard worker. He wanted to build a house that was big and strong, so he built his house out of bricks. It was a good house with a fireplace and chimney. Then he sat back, relaxed, and sang himself a little song.

The next day, a big, bad wolf walked by the lane where the three little pigs lived. He saw the straw house, and he smelled the little pig inside. So he knocked on the door to the straw house and said, "Little pig, little pig, let me come in!"

But the little pig saw the big, bad wolf's furry paws through the keyhole, so he replied, "Not by the hairs of my chinny-chin-chin!"

The big, bad wolf said, "Then I'll huff, and I'll puff, and I'll blow your house down!"

But before he could, Wentaway came down and blew all the houses away.

Goldyn's *New Poems for Children*

Chapter 7
Gwen

Gwen knew the moment her hotel monitor chimed that the person on the other end would be Carter Rhodes. He was a man on a mission; he was not one to let anything sit for long. She had been awake for exactly forty minutes, had just finished breakfast in her bathrobe and nothing else, and with her hair looking like a willow after a hurricane, she wasn't about to enable her camera. Rhodes, on the other hand, looked as if he'd had a film crew arranging his wardrobe and hair for the past hour.

"Dr. Baré—Gwen. Shall I address you as 'Your Highness'?"

She rubbed her temples. "Not feeling very royal this morning."

"I trust you slept well?"

"Very well, thank you. Just a case of rocket lag."

"And did you try the *bolo de fuba*?"

The smell of baked cornmeal still hung in her room, the taste of cream lingering on her tongue. "Scrumptious."

"Told you. Now, if you can have your affairs in order by October first, we'll get you on the shuttle to Phobos. You're welcome to wait here, but I imagine you have things to do back home in the interim."

"Yeah, must tend to the little beasties in my McMurdo vats. So is there much of a layover at Phobos?"

"No, we've timed things so you're only there for a few days before you transfer to a cycling cruiser out to rendezvous with Iapetus. You'll take a short excursion vehicle to Enceladus. Naturally, there will be no greeting party."

Gwen thought about the sun high in the Venusian sky, and how tiny it would look in the outer system, shriveled to a brilliant mustard seed. She thought of jumbled Venusian basalt glowering under a sulfuric acid haze, and of the burning cold of Enceladus ices, blues and greys and white, under a snowy dusting from the gossamer geysers. She wasn't sure which was more foreboding.

"You'll need help getting below the ice, obviously," Carter said.

"Obviously."

"We've arranged for you to meet someone at Phobos Station. His name's Antonio Vincenzi. He's one of our contractors; really sharp guy from the

University of Padua. English is his third or fourth language, and sometimes he sounds a little entertaining, but don't let it fool you. The man's a genius at what he does." A photo appeared on the split screen. "Tony worked with key personnel at both Enceladus centers—back when there was somebody to work with—and he's made quite a few waves observing the interface between the ice and ocean. We've used him in the role of geologist. His PhD is in planetary glaciology. If there's anything left down there, he's your man to play tour guide. He can help you get under the ice at the various entry interfaces."

"That will be good, thank you."

Carter leaned toward the screen confidentially. "He's pretty good about knowing where to get through the ice, but it's what's underneath that concerns us. I'm anxious to see what you guys find."

"Me, too," she said. *You have no idea.*

"Just one more thing," he said, glancing down with a sheepish look. "I guess I don't have to tell you that we're on a timeline here. We don't know how long the eastern grid can last in its weakened state, and if it goes, we start losing people again. Losing the ones in the north to cold, and the ones in the south to heat waves, losing medical centers and communications. You know the drill. If we can get these power sources from Enceladus, we just may get on top of it."

"That would be my vote, too," she said.

"And there could be complicating factors." He furrowed his brow. *Why didn't he just come out and say it?*

"Like what?" Gwen said, hoping to nudge him into whatever it was he was putting off.

He studied his steepled hands, looking like a penitent at an altar, and then brought his gaze back up. "There's a chance—just a slim chance, you understand—that the power is still flowing because somebody ended up there and is using it."

"Somebody like who?"

He shrugged. "There are lots of possibilities. Wentaway shuffled things around. People. Social structure. Trade. Transportation. In the aftermath, a lot of things went missing. While the normal channels got interrupted, quite a few people fled the Martian settlements, and some never came back from Miranda or Triton."

"Yes, Triton was a happening place, from what I've read."

"Downright cosmopolitan. They gave up a lot to leave, but some stayed. Who's to say they didn't migrate to Enceladus before all the dust settled? Fired up the reactors again and made a new, quiet home for themselves?"

Gwen scratched her chin dubiously. "Triton? I don't know…"

"Triton's one of many possibilities. There were a slew of survivors from Ariel Station who never made it back to Mars. They just never arrived. The point is, if anybody returned to homestead on Enceladus, they're not going to like you taking the extant power sources."

"I see. We'd have to negotiate with them, I suppose. Then it becomes a diplomatic mission."

"Or a military one. Much more efficient. We can't lose focus on what this is all about. If somebody's there, it's the needs of a tiny renegade colony versus the critical needs of an entire planet. The home world."

Gwen swallowed hard. She was beginning to understand the full extent of Carter Rhodes' tenacity, his single-mindedness.. Undoubtedly that was why the World Council chose him to head up the initiative for getting Earth back on its feet. But his laser-focus scared her.

He was talking again. "I'm sending you the travel details now. Tony can be at Phobos in a couple weeks. It's short notice, and it's a quick turnaround for you to get home and back on the road again. We hope you can make that work."

The pace of Gwen's life was about to change from comfortable hopping among four Venus stations to break-neck planet vaulting.

Enceladus.

Anywhere but Enceladus.

***

Gwen showered. She ran the water as hot as it would go, trying to wash something away that wouldn't rinse off. As she dressed, she noticed the message light blinking away on her monitor. The message was marked "confidential." It carried an identifier she didn't recognize: i7p.

"So much secrecy," she clucked her tongue. "So much drama." She tapped the screen. The message contained no audio/visual, only text:

Gwendolyn, my dear. It has been too long. As you probably realize by now, I've been incognito by design. I like the distance and the lack of supervision, rebel that I am. Now I see that you may be planning a trip of your own out to my neighborhood. I would be happy to show you—and you alone—around my laboratories on Iapetus. But if your plans include a stopover at Enceladus, please change them. Enceladus is recovering. The place needs time to heal. Don't come, and don't let them come, whoever it is you are working for. From what I hear, Carter Rhodes is a good man, but he represents dangerous forces. Please contact me at the noted exchange above. I monitor this feed a couple times a week. I need to know what is going on, Gwen. We can talk more if you want. One more thing: Sophia and Hugh would be very proud of you.

Your Melciéna

Gwen stared at the screen, her jaw slack and breath paralyzed. So much lay in the message. The mysterious Melciéna Valentine had resurfaced. For her. And while many had guessed where she might be, now Gwen knew that Melciéna was somewhere near Saturn. i7p. Iapetus? She reread the message. The sender had been careful not to say she was actually on Iapetus, but rather "in the neighborhood."

What bothered her was how much the woman seemed to know. She knew Gwen was on her way to the outer system, despite the fact that she had spent essentially all of her post-college life in the terrestrials. She knew of Gwen's involvement with

Carter Rhodes, and she guessed, at least, that her voyage had something to do with Enceladus. Despite her invisibility, Melciéna clearly still had her finger on the pulse of the solar system. Still the magician.

And what was this business about Enceladus recovering? Recovering from what? How does an abandoned settlement recover from anything? Perhaps, as Carter Rhodes had posited, there were bands of survivors from other settlements hiding away on the seafloor of the tiny ice moon, building back the infrastructure into a viable settlement.

But it was the last sentence that caught her short. *Sophia and Hugh would be proud of you.* She supposed Melciéna was attempting to be kind. She was probably still trying to heal Gwen, absolve her, from a distance. Ever since the death of Gwen's father, Melciéna had been more attentive, right up until the final abandonment. As human civilization left Enceladus, it took its warmth with it, and the little moon reverted to a silent, icy globe. And as it cooled, so did Gwen's relationship with Melciéna. Her gradual disappearance helped Gwen to forget. Now, she was back, inserting herself into Gwen's life where she wasn't wanted.

The idea soured her lunch, her last meal on Earth for a very long time.

Chapter 8
Chevalier

Adjutant Chevalier enjoyed the relaxed environment engendered by Carter Rhodes. The Colonel set an informal tone within the office, and the atmosphere made for efficient work and high morale. But this morning's summons was not so relaxed. It had all the informality of a Courts Martial.

Chevalier knocked once on the inner doors and entered. Rhodes sat before a triad of consoles, all rising from the desktop like a small cityscape.

"Sir?"

Rhodes rested his elbows on the table, propping his chin up with his fists. "I've been thinking about all sorts of things. Enceladus, mostly, and what Gwen Baré might find there."

"Yes?"

"Yes indeed," Rhodes said, his monitors sinking back into the tabletop. "I think we need to have a Plan B."

"Why? We know there's at least *some* power there, and roughly what kind."

"Not that. Baré is a capable analyst. She'll know how best to utilize whatever energy sources are still intact there."

"Then what's the concern?"

"It's what she might find there, Armand. We want nothing to get in the way of our acquiring any usable power from the Enceladus settlements, no matter what may await us on the ocean floor."

"Or *who* is waiting there. Isn't that what you're really getting at?"

"Precisely."

Chevalier frowned. "I'm still not following. The settlements were abandoned outright, with the only unknown being the cofounder–"

"Melciéna Valentine, yes," Rhodes filled in the blank.

"And the sub-human slaves were undoubtedly left out in the cold, so to speak. No chance of survival. Not after all this time."

Rhodes' eyes flared. "Not slaves, Armand. They were never slaves. They were more like biological automatons. No consciousness. Nothing like humans. And illegal as hell. Cloned by a bunch of renegade scientists who thought they could get away with it. Valentine was sent in to fix what they had done. And that's another story. But what if the missing survivors from Ariel Station ended up there, maybe set up shop?" He began to wave his hands around like a symphony conductor. "Or the odd band of Martian refugees?"

"That's a long way to go for some peace and quiet," Chevalier objected.

"Nonsense. Think Australia. After twenty-five years, people might get territorial. Valentine is still out there, and who knows what she's been up to? She could have given aid to any group that might have tried to move in."

"Sure, if it was in her best interest. The woman doesn't strike me as an altruistic Florence Nightingale type."

Rhodes leaned forward. "If anybody, *anybody* decided to homestead in the abandoned settlements down there, they won't be wanting to give up any of their power supplies, whether that's nukes or water current generators or bioreactors or whatever else might still be operational. Countless things could have happened under the ice. We don't know what, and that's why they call it a 'Plan B,' right?"

"I suppose so." Chevalier sounded dubious.

"All right. I want you to organize a military sortie. Have a full platoon at Iapetus ready to board a transport as soon as the others report back."

"Our resources at Iapetus are spread thin, sir. Even if we have all hands on deck, it may only amount to half a company. Maybe sixty personnel."

"Then we back it up from Earth. There will still be a decent launch window if we use one of the more advanced propulsion jobs; a Viper transport should do the trick. When Baré and Vincenzi file their analysis, we should be better able to make a decision on whether to move and how much firepower we need to bring."

"Military action at Enceladus. We'll have to stage everything through Iapetus. The logistics will be a nightmare." Chevalier said, as if warming to the idea.

"That's why we prepare ahead. And remember, it's just a contingency. Baré's vehicle will be capable of transporting just about anything back here that we need. And Baré is our best resource for evaluating what energy sources will best serve us. But at this late date, we don't want anyone or anything getting in the way of a timely retrieval."

"Plan B. Got it." He lowered his voice. "Do you have any reason—I mean specifically—to think we might need to take the reactors by force?"

Rhodes steepled his hands. "Let's just say that the whole operation—the entire history of the place—makes me nervous as hell. It's remote. Remote operations never go as planned, and distances make it difficult to react efficiently."

"I hear we've got engineers working on something that's a lot faster," Chevalier said.

"Years away, Armand. Years. Certainly nothing anybody can wait for at this point. Meantime, recon should arrive at Enceladus very soon. I'm hoping we'll get some better intelligence than we have from our probes."

"When are they due?"

"They achieved Saturn orbit late last night, so they should be in orbit around Enceladus tomorrow morning."

By the afternoon, Chevalier had selected a group of highly trained Special Ops personnel to be culled from terrestrial sources. He did not alert any of the team, reasoning that they would not need to know of their special status unless, and until, action was required. Special Ops were used to working that way. Chevalier then contacted Iapetus control and raised them to orange level.

Chevalier sat at his desk, eyes closed, head resting against his chair back. It had been a surprisingly straightforward job: the organization of infrastructure and the projection of forces. So why did he feel so uneasy about it?

***

Captain Martin Davis was a career special forces member, a supreme strategist, and a damned good pilot. He was more than happy to endure the hardships of a cramped craft with a souped-up drive like this one. The ship had the newest sensor array, more advanced than anything already stationed on Iapetus. Hell, the Iapetus base was hardly more than a glorified truck stop.

Yes, Saturn was a long way, and the little ship was short on luxuries, but Davis was as driven as Colonel Rhodes, and he would make his superior proud. There might even be a promotion in it. After a trip like this, with these whiney recruits, there oughta be.

"Coming up on four," the copilot called out.

"Roger that," Davis said, watching the trajectory data feed. They were just passing Tethys now, spiraling in toward their icy destination. The light of Saturn bathed the backside of the rugged moon in golden warmth. Haloes around ray craters glistened in coppery countenance. Up ahead, the phantom glow of the E-ring spread a brilliant line across the planet's face. No time to enjoy the scenery, though. There would be later. His buddies on Iapetus had promised him a tall, dark lager after the mission. For now, it was time for business.

"Comms," Davis barked. "Report."

"Continuing to ping both Atlantis and Thera facilities, sir, as well as the three marked outposts. Still no response. Not even static. Nobody's down there."

"Roger." He leaned toward his copilot and keyed the mic off. "That's good. Makes things a lot easier."

"The place is deader than a doornail."

"Not completely," Davis tapped his screen. "Still getting lots of power output from somewhere down there. There may be no inhabitants, but I'll bet Rhodes will get plenty of his power plants."

"Excellent."

A red light blinked on both cockpit monitors as a faint alarm chimed.

"What the hell?" Davis murmured.

The navigator's voice, a strained contralto, came online. "We've got bogies. Two missiles coming from behind. I'm trying to triangulate now."

"Where'd they come from?" Davis felt the rush of adrenaline, the kind you get only in combat. The furnace in his gut sent heat up his neck, tensed his muscles, focused his senses to a razor's edge.

With no response from navigation, he turned to the copilot. "Countermeasures."

"Aye, sir."

"Navigation, *how close?*"

"Nearest is passing the eight hundred click mark, sir. Still closing. The other is close behind."

"Shit," Davis whispered, opening up communication with Earth. There was no time to wait for security protocols. "This is Captain Davis with coded comm. Colonel Rhodes, the system out here is not secured. I repeat, the Saturn system is not secured. We have two missiles on approach. Source unknown."

Static filled the radio spectrum. He had no idea where it was coming from. It was as if someone was jamming their transmissions.

Davis called the navigator again. "Any luck with that launch location?"

The navigator paused before replying. She was new to battle, but she gave a professional performance. David appreciated how she didn't lose her cool under the pressure. This was better than any simulation they could have given her. "Maybe Phoebe? Source may be above, not on the Saturn side. Iapetus is in the wrong place, but maybe—they're coming right up our tailpipe, sir!"

Davis keyed Iapetus communications. "Iapetus, do you read? Two bogies close on our tail. They seem to be evading our countermeasures. We–"

The first blast interrupted his report. The engine on the port side exploded. The ship began to pirouette, its spin shoving Davis against his seat. He reached for the controls, his arm heavy in the g-load. His vision began to telescope inward. "Goddammit, what about those countermeasures?" He had to get a message to Rhodes.

The second blast hit, but when Davis tried to say something more to Earth, there was nothing left to breathe.

Chapter 9
Gwen–October 2229

"I don't like it." Josh Aotea paced across the small space in Gwen's room, his hands firmly clasped together behind his back. Gwen had deployed the bed from the ceiling to give her luggage a place to sit; there was barely enough room for him to walk between her bed and the wall. "To Earth, then Mars, and then off into the hinterlands. I don't like you going all the way out there, not knowing what your mission is."

Gwen folded a blouse and stuffed it into a bright orange duffel bag. "I didn't say it was a secret to me, silly. I'm just not supposed to talk about it. You're going to wear a trench into the floor. Why don't you light somewhere?"

"I still don't like it. I brought you on board the Venus constellation. I feel…" He stopped pacing for a moment, glancing right and left. "…*responsible* for you."

"Like a goose and her gosling?" she scoffed.

"Like a big brother. Or a really close older second cousin, twice removed."

"That's sweet, but I don't need any more relatives, thank you. Come on, you can pretty much guess what it's about just by what you know about me. Not much mystery there."

"What I know about you is exactly what makes me nervous. When you're not gallivanting through the skies here, you're off on another walkabout between planets. You have a terminal case of the wanderlust, my dear."

She tossed a quantum tablet into the duffel and sealed it. "Guess you can add Enceladus to my list of vacation spots."

"It's going to get you killed one of these days," he grumbled.

"I'm surprised at you, Josh. I thought this would appeal to the administrator in you: following orders, chain of command and all that."

He punched the bridge of his glasses up his nose defiantly. "I'm all for asking questions at appropriate times, too. I always sleep better knowing the borders."

"Not many borders between here and Saturn. Besides, you said you were going to take a trip. You picked out a place?"

Josh looked at the floor. No trench. "I was thinking of the Aegean. Crete, Athens, Knossos, all those classics. But I was hoping to go with someone."

"Sorry it didn't work out with old whatshername, but no reason for you not to go on your own. Traveling solo can be invigorating. You've been stuck here for too long. You need a break." She grinned. "Listen to me; now who's being mother goose?"

"I could wait."

"For what?"

"For you to get back."

Every sound in the room became a stark solo: the fan of the environmental control system, the gentle ticking of the wall electronics, her music background. The playful note in his voice seemed forced. She wasn't buying it. Gwen reached across the bed and put a hand on his shoulder. "Josh, we're good together as it is. You and I, we're a great team and we do important things for the colonies. Let's not screw it up. Okay?"

"Sure, 'course not. I just thought it might be fun. Just an afterthought, really. I can't breathe."

He leaned against the wall.

Gwen stepped over to him. "You really can't breathe?"

He nodded, his chest heaving.

"Here, lean over," Gwen said. "Head between your knees. Take in a deep breath."

"Sure," he wheezed.

Gwen stepped to the counter, grabbed a small packing bag, and handed it to him. "Breathe into this. It will help."

He nodded again, blowing into the bag. In a few moments, the color had returned to his cheeks and his breathing relaxed.

"Thanks, Gwen," he said into the bag.

She ushered him to the edge of the bed and sat him down. "There. Just relax."

"The last time this happened was when I cut my finger. Don't do well with blood, you know."

"Or with bad news, I guess."

He smiled self-consciously. "I suppose not. You're right, of course. Business and romance. Never a good combination." He put the bag down on the bed. "Maybe I'm crazy. All that stuff I said in the restaurant? About you being driven? I think it's a little different. Gwen, your heart is as big around as a gas giant. If you let in all the stuff that's come before, all the love and the anger and the triumph and the loss, it would kill you. The only way you can protect that heart of yours is to keep those things at bay, a little distance away. I think that's why you stay up late at night with your microbial beasties. Plus, of course, you really do love your work."

He craned his neck to look out the window. Night had fallen, but the distant clouds sparkled with lightning. On the horizon, the dull orange glow of an active volcano painted the sky, merging in the north with a band of purple, the delineation between world and twilight sky. A few stars twinkled above the brooding cloudscape. "It's a long way out there. Lotta miles between us and the lord of the rings."

Gwen forced a casual tone. "Colonel Rhodes and his people are helping me get my ducks in order. Or should I say geese? That's not what's worrying me."

Josh turned to meet her eyes. "What is, then?"

9 Gwen–October 2229

She held up an empty wrapper. "I hear there is a critical shortage of sunflower seeds beyond the orbit of Mars."

Josh scratched his chin, frowning dramatically. "That could seriously affect your work."

***

He saw her off at the launch port the next evening. The sleek hypersonic transport came to visit Mac IV twice each terrestrial month. The spear of gold, silver and black hung vertically, docked to the side of Mac IV's main landing platform, its outlines fading in and out against the backdrop of nocturnal clouds. Spotlights illuminated the side, creating the impression of a long train of glowing ovals floating in the darkness. Beams from the station's spotlights washed across a few areas on the ship's flank, tying some of the random white ellipses together, giving them concrete form against the diaphanous world beyond.

As Gwen headed toward the gangway, Josh grabbed her gently by the elbow. "This should last you awhile." he said, handing her a football-sized package.

She opened the flap and peeked inside. "Sunflower seeds? Where did you get so many?"

He started back down the corridor, calling over his shoulder, "They say they come from actual sunflowers. I'm no botanist."

"Thanks, Josh!" She had no idea where he could have gotten such a haul. He must have planned ahead. How sweet.

Josh disappeared into a small crowd of milling passengers and well-wishers. Gwen turned to the entry, scanned her ticket, and entered the airlock with a trio of others. Doors sealed behind, and opened in front. They entered the craft itself, taking their places behind a short line. As they did, four more came through the doors.

Gwen passed her hand over the biometric scanner and continued into the main cabin. The entire craft was sideways, with seats on one side and ceiling on the other, dropping straight down like an elevator shaft. Rails extended from the walls, suspending a small platform between, just at the center of the shaft that would become the central aisle. Gwen and her three cohorts stepped aboard. The platform descended gracefully, sliding down to the level of their seats. She waited for the young man to take his window seat, and then she slid her bag into the overhead and took her seat, lying back into it and letting her head fall against the headrest. The platform ascended again, soon returning to pass by toward the seats on the next level below. Gwen glanced around. The ceiling was low. The twin seats on each side paraded down to a bulkhead in the middeck of the craft, with a secondary airlock separating the two compartments. Silvery gray carpet met almond curved walls. Each row of seats had a small window, much like the old passenger airliners used to have.

Off on another journey. Her friends said they got sick of travel, tired of hotels and exotic food that didn't agree with them. Gwen had not gotten to that place, and hoped she never would. Travel had been a way of life in childhood, and these days, if she didn't make a major move every year or two, she knew she would go crazy.

9 Gwen–October 2229

Her five years among the Venus colonies was a remarkably stable stint, but it had been peppered with trips to Phobos, Mars and Ganymede. Still, the wide-open skies of Venus fed her wanderlust, calmed her spirit.

But now, it was time for a very long trip. All her other voyages had entailed an endpoint she understood. Her parents had carefully explained the family's destinations, or Gwen herself had carried out detailed research of upcoming vacation spots. This time, this trip, felt unsettled. Enceladus was one of those dark places where all communication had fallen silent. Iapetus was nearly as quiet, though it still had a few people. Not so Enceladus. Anybody left in that frigid Davey Jones' Locker was by now a frozen corpse drifting on currents in eternal shadows.

The liner screamed through the night's gossamer clouds. Gwen glanced back down the line of small casements. The lights of McMurdo IV glimmered through the hazes outside, like embers fading and surging in the gusts of a blacksmith's bellows. The buoyant station disappeared and the stars blossomed above.

She had made the trip up to the Venus orbital hub many times, but not for something like this. This voyage was different, something new, a trip far out of her comfort zone. She would stage through the Earth L1 station, steering clear of the home world's gravity well. L1 would be a roadside stop, a quick touch-and-go. And then onward. The very thought made her dizzy.

The Venus cruise had few distractions. The little vessel had two restaurants, a small gym, and a virtual theater that seemed continually full. Most passengers stuck with their own personal devices for entertainment. She tried to reserve a sim theater time slot online. The only one available was a half-hour session two days out from Earth, and that one she'd have to share. She signed in and then caught up on reading and naps. Space travel is like that: long periods of dark nothingness punctuated by short bursts of excitement at the rare arrivals and departures.

The Earth was looking brighter and distinctly blue by the time her session began. The attendant at the entry handed her a headset, sensor boots and gloves, and an electronic mace. She stepped through the hatch into a fifteen by thirty foot room padded on all surfaces with projection skin. As Gwen donned her goggles, rolling hills emerged, with banks of evergreen rearing up in the distance and a waterfall thundering from a nearby cliff. Snow fell from the sky, and the temperature of the room dropped to enhance the experience. A castle loomed atop a cliff in the distance. Yes, a medieval castle, a place of stone and cold and dark windows. If her mission was unsuccessful, most of Europe might just return to a scene very much like this one.

She shared her forested medieval battle with some teens from Mac III, who all appeared in her field of view wearing warrior garb. Stealth was not the approach of the Mac III gang. Their feeble attempts to hide in ambush were thwarted by their giggling and hollering. By session's end, Gwen had lopped off a limb of two different players, the benefit of her years of MMA. For her part, Gwen sustained only bruises, which, as it turned out, were more real than virtual. The cost of digital simulation combat in close quarters. Despite the occasional visual dropouts and objects superimposed over each other, the play was good exercise and better distraction. Still, she would have preferred the Na Pali Surfing package or the Loch Ness Monster sailing theme. Maybe the Mars cruise would have more choices.

9 Gwen–October 2229

She watched the happy little mob scamper down the corridor, back toward the passenger quarters. Such innocence. But for the worlds they knew, the reality they experienced, they were living on borrowed time. Pondering the trends chilled her. *Before you know it, our skein will degrade into an old-style Internet, and from there it's a short hop to parchment and quill pens.*

The day of arrival brought the Earth into focus as a brilliant blue and white pea floating off the starboard side. This was about as close as they would get to the home world; a million miles was a short distance compared to what was ahead. She made her way to the observation deck to see the docking, but the windows were small and the crowds large, so she watched it on her room monitor.

Once docked to the outer ring, the gong sounded and everyone disembarked. Gwen would board something a lot larger for her trip to Mars, and that was fine with her. The Earth/Venus shuttles were more cramped than the luxurious trans-Mars cruisers, and the extra room would be a welcome change.

The station's 1/8 g was about what the acceleration force had been aboard ship, but the floors here were subtly stairstepped to compensate for the coriolis force of the spin. Even her hotel room would have canted floor, walls and ceiling, she knew. The entire effect felt homey, almost natural.

The arrival lounge had the distinct space-station aroma of plastic, ozone and gymnasium. Passengers made their way up various corridors toward the central hotel lobbies and quarters. The Mac III teen team constantly announced their presence with their laughter and pinball-progress along the walls of the corridors. Gwen listened as their chattering faded down one of the walkways. The halls were tube-like, with open-air escalators snaking in all directions past potted plants (probably fake) and projections of smiling tourists (also probably fake). Shopkeepers were just opening their booths for the morning rush, and most of the restaurants sat shuttered. Gwen hated these interplanetary schedules; they always seemed to reach their destination too early in the morning or too late at night for the good restaurants to be open.

She arrived at her assigned lodgings and dropped her bag onto a small table next to the stowed bed area. She noticed her console's message light blinking. She tapped on. The message status flashed "confidential." The face of Carter Rhodes snowed into form.

"Hey Gwen, I assume you have arrived at L1. Hope your trip was comfortable. Sometimes the government dime doesn't pay for the most top-drawer of accommodations, but this ticket looked okay to Chevalier. The Mars leg will be a step up, I'm sure. Now, to business. I wanted to update you on a couple news items. First of all, the Mars council has turned us down again on solar panels. They say they're tapped out, nothing more to give, and their manufacturing is behind even for domestic orders, so Earth comes at the end of the line (there are some politics going on there, I'm sure). This news was made a bit more significant last night. Most of Europe's Atlantic coast dove into darkness at about 8 pm Paris time. We're trying to protect the global grid from a catastrophic rolling failure. This makes your mission all the more imperative. Godspeed to Enceladus, Gwen. I hope you find what we're looking for. Rhodes signing off."

Suddenly, she was a child again, with a child's nightmare dreads. This threat to the world was not some bar chart, not an abstract drop in power levels. It was a creeping darkness, a thing of talons and fangs and cold malevolence. It slithered like a black amoeba, engulfing light and life. And it was out to dull everything noble and good and dazzling. It was the enemy of luminescence, the confederate of the bleak and gloomy.

Nothing like a little more pressure.

***

Rhodes made a fist and jammed it into the palm of his other hand. "Nope. No, she doesn't need to know at this point. Neither of them does. There's no need."

Chavalier grunted. "At least so they know what to look out for, what they might be getting into out there."

"To what purpose? They're not in an armed military transport. They'll be arriving in a tub that can't even avoid running into a blob of rock, let alone an incoming smart bomb. And our military backup will be a couple days late to the party. No, at this point, I say ignorance is bliss."

"They won't be blissful for long if one of those mystery missiles flies up their asses. Sir."

"Which makes it all the more prudent for us to find out the source, and if possible, take it out. But for Iapetus, a preemptive strike is pretty much out of the question. They don't have the resources. Saturn's too far out for our own forces to get there in good time, and our two experts have gotten too much of a head start. They've just got to take their chances and get down through the ice."

Chevalier sighed. "The whole thing makes me feel guilty."

Rhodes squared his shoulders. "Marine!"

Chevalier straightened. "Sir."

Rhodes relaxed his frame. He looked at his adjutant with sympathy, dropping the drill sergeant tone. "Chevalier. In military action, there is no room for guilt. Things must get done. We're not out to spare people's feelings or make our comrades comfy. We have a job to do."

"But sir, to me that's the point. They aren't military. Do we have the right to put them in harm's way just because they volunteered?"

Rhodes stood and crossed to the window. A large orbital drew an incandescent arc across the night skies of Sao Paolo. "Damn, Armand, I do feel responsible. But there's not a lot we can do to help from here. Let's let them concentrate so they get done and return for home, for safe harbor, sooner." He tapped the tabletop. A screen rose up and began to scroll data. "Any news on the trajectory?"

Chevalier pointed toward the screen. "The latest models indicate that it could not possibly have come from Titan, and with our facilities there, Iapetus is also unlikely. That leaves us with an impossibility."

"Enceladus," Rhodes said. He groaned. "It's all dead down there. Who could be trying to warn us off of that ice ball? Who would want to? There's nothing to defend but empty cities and icewater seas. This could be very bad. If somebody is there, laying claim to things…"

"Whoever it is, they can't possibly need every reactor on the moon. We could, you know, negotiate."

"Exactly what I want to avoid. Time is of the essence, and we've got no extra hours to beg for power supplies. This is bad. This is looking more and more like steps must be taken. Your special ops are in place?"

"Ready to leave whenever you give the word."

"We may just need our Plan B sooner than I thought."

Chapter 10
Antonio Vincenzi

So far, Antonio Vincenzi's mission had been frustrating as hell. They'd given him very little information to work with, and even less about the mission's genesis. To top it off, Phobos itself was a frustrating place. Mars loomed over the horizon as a constant presence, a great, baleful dome haunting every canyon and crater on this side of the big rock. The red orb was enough to make anyone nervous. But on Phobos, a nervous person could not pace. The least movement brought both feet off the floor. Had he wanted to go outside, he could jump a thousand times as high as he could on Earth. Imagine the hang time! Half an hour? He didn't want to go outside. The magboots helped, but they made anything like pacing completely artificial and unfulfilling. So he leaned against the bulkhead next to the panoramic window, letting his body settle against wall and floor, looking out at the coal-black wall of Roche Crater with the arc of Mars eating away at the black sky from behind, watching the skies for this energy guru he was supposed to spend the next half year with. What would she be like? From Venus, yes, but Gwen Baré had American roots, and Americans could be abrasive. He hoped she would at least be civil.

Against the velvet black, through the diaphanous dome crowning Roche, he saw her coming: a flash of silver and bronze moving across the stars. The spurt of the little attitude control jets brought the glimmering spot around, twirled it into something else. Now he could see the gangly shape of an interplanetary cruiser, not sleek or ship-like, but more a pile of construction materials. These kinds of vehicles never saw an atmosphere, so they needn't be sleek. But couldn't they at least have some kind of respectable symmetry or organization?

***

Gwen peered down at the approaching landscape, battered and vertiginous. Lofty cliffs met the sheer walls of craters and chasms, black shadows laid against molten glass highlights. Strange how a lump of stone as dark as charcoal could look so bright. The transport headed north, toward the pole. Directly below, a dome spanned the top of a crater. Next to it, the docking tower extended up toward them, an Eiffel against a velvet sky. Unseen radar and flashing lights brought the ship home. It docked against the tower, a structure that seemed far too flimsy to hold a vessel like this. Gwen was used to the rugged pillars and guy wires of the McMurdo cloud cities, built to withstand the gravity and winds of Venus. This edifice looked downright precarious. But in the feeble gravity of Phobos, the great mast was plenty resilient. The massive cruise ship docked to the side of the gossamer tower and clung on to two sets of davits for dear life, sending a subtle wave through the tower to the ground and back again before it settled enough for its passengers to disembark.

As she unstrapped from her couch, a corpulent man next to her said, "Know what I love about Phobos? The gravity. It's my second time here. Know how much I weigh?"

Gwen smiled noncommittally.

He leaned toward her and bellowed, "Four ounces. Can you imagine? Four. Friggin. Ounces!" His laughter sent ripples through his jowls as his face reddened. "Not only that, but our baggage is essentially weightless, so the port is more easy to negotiate than Mars or Earth or somewhere like that. It's amazing. Wow."

He grabbed his personal case, slipped his tablet into it, snapped it shut and saluted her, disappearing down the gangway.

Gwen followed the flow of passengers as they float-walked down the tunnel, descending effortlessly toward the gunpowder moon. By the time she reached the end of the tunnel, she had checked into her room remotely on her personal comm. The tower foundation opened into a wide lounge with panoramic windows displaying the bruised landscape beyond. The spaceport jutted from the sunny side of the crater rim, but the far wall of the crater—in shadow—was bathed in orange light. Somewhere behind them, Mars loomed large.

The receiving area had a different feel than the interplanetary port in Sao Paulo. People spoke in hushed voices. Every wall had the same padded carpeting as the floor and ceiling, and chairs extended from several angles. A ring of imaging surfaces displayed scenes of jungle, ocean and mountains, with background audio to match. Crickets chirped, breezes whispered through trees, birds sang. In fact, Gwen realized that the place was running the same soundtrack she had heard in the main spaceport lobby down on Mars; she recognized the *ker-REER, ker-REER* of a jungle bird, followed by thunder. Maybe they were on a budget.

Several passengers had strapped in around small tables, where they sipped piping beverages through straws. Gwen looked up at the wall menu next to a small galley.

"May I get you something, Dr. Baré?" The voice from behind startled her. She pivoted, drifting off the surface she had come to rest upon.

"Dr. Vincenzi?"

"How about Tony? I work better if I'm not under pressure. 'Doctor' sounds too much like I know what I'm doing."

She laughed. "I've been told that's the case, at least where Enceladus is concerned."

Gwen considered her assigned companion. Antonio Vincenzi was a small, entertaining man. He had the rich skin of Mediterranean climes, and she saw fire in his gaze. His eyes reminded her of two little animals scurrying around inside dark caverns. He had tied his luxuriant, waved hair into a salt-and-pepper stream behind his head, a ponytail that sometimes drifted from behind him in the low gravity.

"I can get you down through the ice, yes. Easy as shooting pie in a barrel." The man spoke like an automatic assault weapon. "And I'm anxious to explore the archaeology. Abandoned outpost, social experiment with genetic engineering, leftovers of a lost civilization. All sounds pretty entertaining to me. And if the government wants to pay our way, it's good with me, signorina. A ticket to Enceladus costs an arm and a foot. Espresso?"

"Yes please."

He inhaled a breath. She took the opportunity to jump in. "So, Tony, what's the plan? When do we catch the cycler?"

"It is only a few days out, they said. We have to leave before it gets here, to catch up speed and meet it as it whizzes by. So tomorrow afternoon, I think, is the shuttle. I saw it. It's a cute little thing. Thank God we won't have to be in it for a long time. Just hours."

Tony retrieved her drink from the galleybot and they settled into seats at a small table. Gwen strapped in, but Tony seemed to enjoy his buoyancy. Every time he gestured, his buttocks left the seat for a moment.

"You are energy guru, yes?"

"That's my area of focus. Bioreactors, mostly."

"Ah yes, the tiny beasts in big drums."

"You might say." Gwen took a sip before she continued. "And what about you? The ice man cometh."

Tony seemed shy to talk about himself. "It is a little odd, actually. My family is in the business of smelting ovens for the asteroid mining industry. For several generations now. Italy was the big pioneer in mining, especially with Pallas and Metis. So my sister and brothers, father and mother, and grandmother all followed in the same footprints. Footsteps. Then came me, the glaciologist." He grinned wide, teeth sparkling out from his rich complexion. "The situation is filled with ironicness."

"Yes, irony."

"Thank you. Yes, please correct when I need it. I like to learn the language better."

"I'm glad you're along. Plenty of flowing ice on Enceladus."

"Sure, but it is not alone. I cut my fangs on the ices of Europa and Ganymede. Europa has a much more dynamic surface, of course. Ganymede is just weird in places. I like weird. Weird makes us use our mind."

"Is Enceladus going to be tricky?"

He frowned. "Not too…tricky. Dynamic flow here and there. Some drift of crust from south to north. We have two provinces close to equator, and those preserve ancient volcanic terrains, but ice flows stopped there a long time ago. No worries. Where we need to go, Atlantis and Thera, no worries."

Gwen stirred her sealed cup with the insert. "But once we're there, now that the place is shut down, how do we get below?"

"Used to be much easier. Powered airlocks. Flexible elevators all the way down to the seafloor. Good old days. But now?" He shrugged dramatically, sticking out his lower lip. "Now we locate the right airlock near Thera. We need primario—the big one—if we're going to get your reactors out."

"The microbial fuel cells can be broken down to fairly small assemblies."

"Yes, but if the nukes still work, they are bigger, yes?"

"If they've got the large ones according to the old specs, the main sources will be big enough that we need to wheel them out."

"So, grande—big—airlock. We power it up using my portable source—should not take much juice—go down through the ice, and hope we don't have to operate in our pressure suits. Work is so slow that way. Like playing Chess with oven mitts."

"That it is," she said. "Sounds reasonable."

It all sounded reasonable, in fact. That was the problem. There was nothing reasonable about this trip. In the *what could possibly go wrong?* department, Gwen knew schedules out here were tenuous. The situation back home was precarious. Each leg of her trip was a link in a feeble chain, and the farther out one went from the inner system, the weaker those links became; the fewer options one had. Carter Rhodes was counting on her, and behind him, there were continents full of people starving—literally—for power.

***

In the wrong kind of crowd, a few months out to Saturn on a leisurely cycling ship could feel like an eternity. Gwen knew from experience. But Tony was quick with a smile, brimming with energy, seemingly eager to help. Perhaps it would not be such a long trip after all.

His voice came to her above the symphony of the busy morning rush. "Amica! I got us a breakfast desk."

"Table," she corrected, sliding into a chair and strapping in for support.

"Table. Si, si. Did you get the itinerary from Colonel Rhodes? For our fate today?"

She massaged her temples and closed her eyes. "I'm sure I did. Haven't checked yet."

"So if we catch the ship tomorrow, we take a head start today."

"What time?"

"Boarding in an hour or so. We must put the 'fast' in 'Breakfast.' We rush to go on a very long and leisurely trip."

"You are a man of irony, aren't you?"

"What can I say? It is the Italian way, no?"

"I doubt that we'll have much company. Nobody goes to the outer system these days."

"We shall see. We shall see. Time will say. There is lots of traffic to the asteroids lately."

"Mining's picked up."

Breakfast with Antonio Vincenzi was starting to strike Gwen as more of a rushed snack. She wondered how efficient her colleague really was. "Are you a fast packer?"

"I am packed already. I've been told I am a very organized person."

"I could learn a lot from you. I'm still tossing things into my bag. You said your room is ground floor?"

"Yes, with a beautiful view of the crater wall. Lots of stones."

"Want to grab a nice panorama of Mars? I'm up on seven."

"Certamente!" he crowed. Tony seemed to do everything with enthusiasm.

The elevator ride provided the first substantial gravity they had felt since arriving on Phobos. The car deposited them on the seventh floor.

"What, did they not give you a hot tub and balcony like mine?" Tony deadpanned.

"Those are on my second floor," Gwen joked. Tony's comment reminded them both just how cramped the quarters were, and with the bed folded into the ceiling, there was only one place to sit. Neither of them took the chair.

Gwen's monitor displayed rolling images from her current home in the Venusian clouds. A small book made of real pulp lay next to it. Tony leaned toward the monitor and gazed at the billowing yellow puffs of a Venus morning. "I have only been to Venus once."

"What did you think of it?"

"Too windy for me, but the skies are beautiful once you get above the fog." He stood away from the monitor too quickly, took a moment to regain his balance, and settled back to the floor.

Gwen sealed her duffel and crossed to the single shuttered window on her wall. She opened the blinds. The orange light of Mars poured in, warming the room. The two travelers gazed down upon the god of war. Morning was just coming to Elysium, the not-as-famous volcanic province on the red planet's side opposite Tharsis, a rise hosting the largest volcanoes in the solar system. Elysium's smaller cone-shaped mountains cast long shadows through the purple mists, disappearing into the night's darkness. A few city lights scattered over the darkened landscape, a rebellion against the dark ages of Wentaway. Gwen pondered the sparkling yellow pinpricks. Life arose again. Light prevailed against the gloom. But it was a battle. And on Earth, that battle was a losing one. No matter: they would take themselves to the front of that war, stare it in the face, and shine what light they could into the black onslaught or die trying. She could imagine Tony and herself encased in the frozen wasteland of the glistening moon. A girl from Venus and a guy from the Mediterranean turned to human popsicles. How was that for irony?

Chapter 11
Circles Within Circles

The waiting area was empty, save for one woman wearing sunglasses and a trench coat, like a femme fatale out of some classic holovid. She even had the hat, out of place in a line where passengers had to don pressure helmets now and then. A flash of blonde under the hat brim swept back, disappearing behind her diamond-studded ears.

Tony had not yet arrived. Gwen drifted over to the woman.

"So…not many of us headed to the outer system," she chirped.

The woman nodded. At first Gwen thought the mystery woman was going to ignore her, but after a pause, she commented, "Not much to see out there anymore. Business?" She gestured at Gwen.

"Me? Yeah, business. Long trip. How about you? Going to Saturn?"

"Further. Business for me, too. Excuse me." The woman tapped her earbud and began to speak in hushed tones. Conversation over.

Tony arrived and checked in. In minutes, the three travelers sat side by side, braving the abrupt acceleration of a craft catching up to a transport. The little ship struggled to meet the passing vessel, making several violent adjustments to its course and acceleration over the next few hours. Finally, the great cycler came into view. Even from a distance, they could see it slowly cartwheeling across the sky, a titanic "Y" structure spinning to create gravity an eighth as strong as Earth's, equivalent to the Moon or Titan.

"I would hate to be the pilot, trying to dock with all that turning metal," Tony said.

"On the other hand," Gwen said, "there's no acceleration to deal with. I imagine docking with a coasting ship is far easier than something more active. Just hop on and wait for your drop-off point on the big circuit. Around and around."

"It looks delicate," Tony said skeptically.

"Doesn't need to hold up too much in zero-g. The slow spin lets all the interior structure be fairly light." She grinned. "Don't be nervous."

The cycler had been looping between Mars and Saturn long enough to get ragged around the edges. Its course had been determined before Wentaway, back when cyclers and cruise ships departed weekly with supplies and tourists. *Those were the days.*

The personal quarters were cramped, the décor outdated, the handholds worn. Mold clung to the panes of the windows. The interior smelled of stale food and disinfectant. It was a big ship, an old-style cruise liner with too much room to fill and too few passengers to do the job. The dimly lit corridors echoed with their passing, the floors creaking as they made their way back to the communal lounge. With each lazy turn of the ship, the walls groaned under the centrifugal forces, threatening to fly apart and end the cruiser's sorry existence. Small LEDs cast long silhouettes across floor and ceiling, a film noir of light and shadow. In the low gravity, the ethereal shapes gave Gwen vertigo. Where was the soothing background soundtrack? Where the mood music and jovial passengers?

"The place could use some atmosphere," she said.

Tony took a sniff. "I'm breathing okay."

"I mean ambience. A lighter disposition."

"Indeed. It might lighten things up to scrape off the scum here and there." He brushed his finger along the pane of a little porthole and grimaced.

"At least we're on our way."

***

The ship had its theaters and spa and workout rooms and restaurants, all with somewhat dated decor. Ultimately, the eateries became the easiest way for Tony and Gwen to pass the time. The two sat at a small table in a fairly bland space called the *International Lounge.* Gwen surveyed the lounge; only four other passengers inhabited the hall. Two young business partners wore the uniforms of miners from the asteroid belt. They were probably bound for a drop-off at Ceres or Vesta as the ship cruised through the belt, but their tabletop monitor displayed split-screen views of several Saturnian moons. Why they wanted to go to Saturn was anyone's guess. Maybe it was another symptom of people's desire to bring back commerce and community to the outer system. She felt like walking over to the table, taking one of them by the pressure collar, and hollering, *What are you thinking? Why aren't you putting your energies into the home planet? Have you thought about what your resources could do for our starving world?* But it was no use. Some couldn't see it, or refused to. Others hadn't been exposed to it, spending all their time elsewhere, like Mars, where the infrastructure was largely unaffected.

The other two were a couple so in love that they seemed to generate their own isolated biome, as if only decompression of the entire vessel would bring them back to reality. *Young love.*

Gwen turned back to her companion. "Antonio, it's been three weeks. We've put it off long enough."

Tony glowered at the tabletop. "I hate these ship tours on little screens."

"It probably only lasts a few minutes, and there might be something we're missing."

"I suppose you're right. I give up." Tony tapped a button. The screen in the tabletop flashed to life. Now, finally, came the mood music.

"Welcome to the Loop. Our cycling transports will take you to where you need to go in comfort and luxury. Visit our three restaurants and bakery deli. Our crown room has human attendants five nights a week to enhance your dining pleasure, and a human chef oversees all of our auto-servers. For your reading pleasure, our full-service electronic library has millions of titles. On the lighter side, work out in our virtual arena. Climb a mountain. Explore the oceans of Europa or the volcanoes of Fiji. Be the queen in our famous life-sized chess tournament. Take part in a bank robbery or a heroic rescue from a high-rise fire. Save a small Duchy from an international spy ring. You be the hero, wherever and whenever you want."

"Speaking of spies," Tony said, glancing toward the hatch at the back of the lounge. The mystery woman had just entered. Her sunglasses were gone, but her downcast eyes volunteered no secrets. The hat was gone, too, replaced by a cascade of platinum blonde hair. She took a table across the room, putting space between herself and the rest of the passengers. Her body language exuded one unmistakable message: do not disturb.

Tony fidgeted. "It must be time to eat again, but maybe somewhere else. What else is there to do on this tub?"

Gwen tapped the screen. "You haven't been listening! There's lots."

"I say we visit the deli. They have donuts today. I saw."

"Let's let our poor little screen finish our tour first."

The cheery voice on the screen tried its best to infuse the room with some energy. "Visit our real swimming pool for a therapeutic dip. Daily events are posted on the–"

A buzzing sound cut the narrative. Suddenly, the screen flashed red and the voice went silent.

Tony stood. "Guess that's the end of our tour. Donuts?"

But Gwen was staring at the lights in the corridor. They matched the red tint on the screen. "Something's up."

The mystery woman had disappeared. The business couple seemed to cower at their corner table. A steward entered the room. "May I have your attention, please?" he said as if addressing an auditorium. His voiced bounced off the walls. "We have suffered a minor technical glitch. We will keep everyone posted as to our repairs progress, but our projections suggest that we may need to forgo our swingby of Vesta. We apologize for any inconvenience."

The announcement clearly agitated the business duo. The steward turned back through the hatch and headed down the corridor.

Gwen stood. "That was damned unsatisfying." She dashed out the doorway in pursuit. Partway down the corridor, she stopped the steward, a slight man with an officious mustache. He wore thick makeup and dark eye shadow, the kind that used to be all the rage for the Ganymede upper crust.

He puffed himself up and said, "May I help you?"

"Just a little clarification," Gwen said innocently. "By forgoing a Vesta swingby, what exactly do you mean? I thought the route just sorta happened."

"The ship will, of course, still follow its trajectory, which in this cycle will take it close to Vesta. But we may not be able to tweak our path to deploy or pick up shuttles."

"What's going on?"

The steward adopted a condescending tone. "My dear, suffice it to say that things are not well with our power supply to the ion engines. Those are what make the ship go where we want it to."

"Are they shunted through the fusion plants?"

"Yes, ma'am, they are," he said in his best you-really-wouldn't-understand voice.

"It seems that the only two problems you could be having are with your power coupling or with the baffle assembly. Everything else is fairly straight-forward hardware."

"Oh," he said uncertainly. "Well, yes, you're probably right. I'll certainly check." He scurried down the hallway.

Tony was standing in the hatch, leaning against the doorjamb with a smile on his face. "That man will not make any more pre-assumptions about you any time soon. Good for you."

Gwen grinned, not bothering to correct him this time.

The steward made another appearance at dinner, in the Crown Room, where it seemed that the balance of passengers had opted to graze. He made a beeline for their table.

"Excuse me, Miss–" he leaned in to read her nametag–"Miss Baré."

Tony sat up. "It is Dr. Baré, please."

"Yes, excuse me Dr. Baré. You were right about the power to the engines. It's something to do with the baffle assembly, which I don't really understand. The captain said you are welcome to come up to the control center and she will give you a tour."

"Thank you very much."

"You are more than welcome." He bowed slightly. His pretentious tone had vanished. "Are you both having a good cruise?"

"Quite," Tony said with the slightest hint of derision. "The service is squisito." The Steward seemed unsure of what the word meant, and excused himself abruptly.

Gwen leaned toward Tony. "He knows Italian about as well as he knows the ship's engines."

***

The control center stood at the hub where the three arms of the ship met. The stars outside spun slowly around a point directly ahead. Captain Haegen introduced herself and her crew. Even in the weightless environment, the captain held a posture that commanded respect. Her close-cropped hair lay against her skull like a helmet, but the practical style seemed fashionable even so. Her shoulder displayed an array of medals, badges and three racks of ribbons. They were all small, not overbearing, but significant. Gwen chided herself for expecting a country bumpkin, a captain who was somehow second-rate on this tub. She was impressed.

The tour was short, as the ship's bridge was simple and cramped. The technology was classical, the interior design bordering on kitsch. As the hub had no gravity, chairs were arranged radially, with everyone's heads pointing toward the center of the room. Gwen always found it disorienting to enter a room in zero g.

"So the bottom line," Gwen said, "is that you can't send power to the engines without a new part that's small and cheap and simple, which you get from either Mars or Earth."

"That's the long and short of it," the Captain said. "We can't figure out a way to jury-rig–"

Gwen held up her hand. "Oh, you don't want to go jury-rigging an ancient reactor like this one. I think fresh parts are your only safe option."

"With your background, I trust your advice," Haegen said. "We had a ship's engineer, but he was a bit green. Didn't apply himself."

One of the crew put in, "He was a slacker."

"It's as hard to get good engineers off-worlds these days as it is those simple parts you mentioned. Mars and Earth are booked up, and the cyclers are low on the totem pole for technical help. I guess everybody figures a ship that spends most of its time coasting doesn't need someone to tend the engines." Haegen looked preoccupied.

Gwen prodded. "There's more, isn't there?"

"These ships are old, but they're robust. Never saw a failure like this."

"Never?"

She shook her head slowly. "This will, of course, put a damper on the plans of passengers and crew from here on out. There's a long loop ahead before we make it to a safe harbor, and Iapetus is the only station left out here that can help restore our provisions. Life on a cycler often entails waiting for the ship to work its way around its path."

Yes, Gwen thought. And Carter Rhodes would not be happy with a multiple-year delay for his little project. But what could anyone do? It looked as though they were trapped on a cycler that could not maneuver, and the loop would take years to transit. Perhaps the only outcome was to ride the ship around the next orbit, hop off at Mars—if that was even possible aboard the crippled ship—and return to her life at Mac IV. There, she would have to wait to regroup, and pray the world's power grid would keep the darkness at bay until she could return to the outer system.

And to the memories.

Deep inside of her, she heard the familiar sound again: the scream of a little girl. "Daddy! No!"

***

Rhodes' message arrived the next morning, while the two sat in the ship's coffee shop. He must have gotten the news about the cycler yesterday. The time delay to Earth now hovered at just under an hour, so he must have been thinking for a while before he hit SEND. Gwen added up several instances in her life where this would have been prudent for her to do. Hindsight.

TO: Baré, Gwendolyn
Vincenzi, Antonio
FROM: Rhodes, Carter
RE: Cycling ship
TIMESTAMP: Nov11/2229/06:31:25GMT

Drs. Baré and Vincenzi: We are examining a handful of options, none of them ideal. There is no obvious way to get you all off of your cycler. The passengers and crew may be retrieved (note that we are not using the term "rescued") by a returning Mars cruiser, but trajectories have yet to be worked out. If we use this option, you and everyone else would stay on the cycler until after it passes through the Saturn system and dives back into the asteroid belt, where you'd be intercepted and ferried back to Mars. This would completely end your mission as it stands now, so we're hoping to come up with an option that would get the two of you off the ship and out to Enceladus as the cycler flies past. If that's not possible, we'll have to let you ride home and send someone else out later. Opportunities to the outer system are few and far between, and this is the least appealing option.

Rhodes had gracefully skipped over another problem at hand: consumables. This cycler relied on resupply at Saturn to make it back to Mars. Iapetus was the only facility beyond the asteroids that had enough inventory to resupply passing ships. But a fast unpiloted supply shuttle didn't have the range to catch the cycler unless the cycler itself could tweak its path. No engines, no tweaking. No tweaking, no food or air. She read further.

We're still looking at the cycler's next waypoint, which will be when you pass close to Iapetus, unless we can dispatch a faster military transport to the asteroid belt. But no transport is going to be available for some time, due to a combination of bad luck, equipment failures, launch windows and politics. Will let you know when we know something more concrete, but unless we find another approach, Enceladus is a no go for you all to depart. Hope you brought a few good books. Rhodes signing off.

Retrieved. What a nice euphemism for bailing out a bunch of people from a gradually unfolding crisis in deep space.

"This is not appealing at all," Tony muttered.

"Which part? There's a lot here that's not appealing."

"The prospect of getting so close to the goal and sailing by Enceladus with only a waving and arrivederci." He stared at the table. "Perhaps Rhodes can send one of the fast military transports to fetch us at Iapetus."

"I doubt they could take the entire passenger and crew manifest from this cruiser. Besides, without a course correction, we'll probably be out of range, even for an unpiloted resupply. Starving to death in deep space. Now *that's* not appealing."

"What about borrowing one of the Galilean cruise ships?"

"They're great between Mars and Jupiter, but not so much beyond. Can't take the cold very well. And they move at a snail's pace."

"Slow trip is better than none at all."

"Depending on how slow, I suppose."

"I am glad you can be so philosophical," Tony said.

He was right. It made her wonder. Why was she not more disappointed? With the high stakes, she should be devastated. But she felt only one surprising emotion: relief.

She glanced up. Tony was smirking. At a time like this.

"What are you looking at?" she snapped.

"You, of course." He paused. He was debating something. Finally, he said, "You are a puzzling one, mi amica."

"Ask me anything."

He tipped his head to the side, a weak smile still there. "You don't want to go, do you?"

Gwen tensed. How much should she say? She dropped her shoulders. "I know what's out there."

"As do I," he said quietly.

"How much time did you spend down there? How close did you get to the Naiads? Close enough to watch them work? Near enough to smell their fish skin or watch them lumber down the hallways like gorillas? Did you look into their eyes and see only black and emptiness?"

Antonio's smile was gone. He looked like a cornered animal. He offered an exaggerated shrug. "I only thought–"

Gwen fought down a shudder. "Look, Tony, it's cold and it's dark and who would want to go out there, anyway?"

"We will bring our own light."

"If we weren't off saving the Earth, this is not a trip I'd sign up for."

"I'm sure Enceladus is short on good restaurants and fornos—bakeries—so I am with you on that count. But the place will be quiet and empty. The only shopping we do is for power plants, and I'm sure we will not be disappointed for those scores. No other shoppers to queue up behind, yes? It is up to us. The ball is in our hoop."

"Assuming we can get there in the first place."

Carter Rhodes had his heroic plan. Gwen had joined in. Now, so had Tony. The plan had been put into action, only to be derailed by a piece of equipment available in most terrestrial hardware stores. To think that such a simple piece of failed technology might kill them before they had the chance to succeed—or fail—at Enceladus!

Tony looked at the grimy porthole and the glistening stars beyond. "Yes, it does seem rather hopeless. These cyclers always resupply as they pass Saturn. If we don't make that connection, we'll all starve by the time we pass through the belt again."

The diminutive Italian was probably right. Rhodes' options were meager and far from optimal, as were Haegen's. No Saturn, no reactors. Nevertheless, this bad news for the people of Earth seemed like respite to Gwen. That thought filled her with guilt. She stood. "I need a serious attitude adjustment. Sorry, Tony. I'm going back to my room for a good sulk. Sort things out."

"Happy sorting," he said cheerfully.

Her quarters felt darker than they had before. Faltering light seemed to be the day's theme. She knew darkness. She had seen it in the eyes of the sea creatures, those humanoids relegated to the depths of an ocean world. Life was not made to live in such places.

Chapter 12
Interval 3 -Tommy- The Recent Past: 2226

"Tommy, this is a big deal. With Mom gone and grandma in and out, your Aunt Rosie is the matriarch. She's the north star of the family, right?"

The boy's chin dipped to his collarbone. He gazed at his toe, digging into the carpet. "Oh, all right. I'll be there as soon as I can," he said before hanging up the call.

Why did dad have to be so overbearing? Life was so hard sometimes. All Tommy needed was another couple hours of *Armies of Treachery* and he would defeat the boss on the next level. There were still a good two hours of public power left, plenty of time for his virtual skills to triumph over his fellow adolescents. But he supposed Dad was right. Aunt Rose had been good to them. She kept things together, even when the lights went out and the heat failed and Mom died. Now the light was going out of her, the heat waning. They could keep her going for another couple years, and maybe even fix her if that doc from Detroit made it out in time. He hated to see all the tubes and wires and hissing machinery. But it was worth the effort: how would he feel if it was his mother or sister or someone, in that cold room?

There were fewer mag rails these days, but he caught one in the nick of time. The cabs rattled along the tracks, past the abandoned buildings and the huddled trios and quartets clustering around the can flames and bonfires. A once-elegant grandfather clock sat upright in a field, blazing away, warming a bunch of what looked like school kids. Further on, a ragged band of hoodlums skulked down a closed-off street, undoubtedly planning something nefarious. Civilization barely holding on. If only the sun would burn through the gray haze.

The buildings of the downtown metropolis, cleaner than the suburbs, reared up along the mag rail route, forests of concrete and steel edifices that somehow remained dark and hulking despite their whitewashed exteriors. Boarded windows gave way to the revitalized city central. He tossed a sideways glance at his place of semi-employment, the library. Windows blazed from within, light showing the way for the dreamers and researchers. With limited power, paper was coming back, often between the covers of books. People were even learning the alphabet again.

12 Interval 3 -Tommy- The Recent Past: 2226

The lights of downtown lent some cheer to the dreary day. The hospital had more than its share of them. Smoke poured from smokestacks at the side of the med center, a rare independent power plant that helped out during the inevitable power disruptions. They said it could take over for three or four hours at a stretch. Judging by the drifting grime, it looked like it was working overtime.

As Tommy approached the front entrance, a loud pop issued from one of the smokestacks. A cloud of sparks billowed out, and then the column of smoke stopped. What was that all about?

He didn't take the elevator. He used the stairs ever since he got stranded in there for a couple hours with some old lady and a screaming baby. Of course Aunt Rosie's room had to be on the seventh floor. Life was hard.

Dad met him in the hallway, all smiles. Tommy tried to control his breathing. He liked taking the stairs two at a time, but it took a moment to recover. Dad ignored the wheeze.

"Great news, son. That guy from Michigan? Dr. Franklin? He's coming to this very hospital next week, and he's bringing a gene therapy that will reverse Rosie's disease. He told me the process is power intensive, but he'll bring his own energy pack so he won't have to rely on the hospital's current. Your Uncle Carter is going to pay for his expenses."

Tommy drummed up a little excitement in his voice. "That's awesome, Dad."

"Yep," he said, looking down the corridor toward Rosie's room. "It'll be great to have her back again. She'll be running circles around us in no time."

The nurse stopped them at the door. "Mr. Belford, it's family only."

"That young man is my fine nephew," came a weak voice from inside.

Tommy followed his father into the room. It smelled of sour sweat and disinfectant, with the slight tang of urine. The patient turned her head feebly, looking across the bed. Her voice sounded like a gentle breeze in the bulrushes of the culvert at home. "Hello boys! Albert, would you hand me my water? Thomas, so glad you came to visit." She held up her hand—her wrist limp—toward him. He took it. Her cool skin felt like parchment, and held the same pallor. He could see the purple veins just under the skin, the tendons pulling like lines inside a camping tent. Some of the veins had kinked into a twisted knot where the nanobots had clogged up. Not a good sign.

"How you doing today, Aunt Rose?"

The lights flickered. Several machines rang out in alarm, singing caution to some remote nurse's station. The room darkened. Only the grey of the lowering sky filtered through the little window. Rosie's eyes rolled back under her eyelids. Her back arched. A loud klaxon howled from a tower of machines in the corner, all attached to Rosie with nasty-looking tubes.

A man in a lab coat stepped through the doorway. "Clear the room," he snapped.

Dad and Tommy paced in the hallway, trying to make sense of the sounds coming from the room. Even the alarms had died now. The emergency lights at the end of the hall seemed to be struggling under the weight of the encroaching darkness. Dad walked to the end of the passage and stood by a floor-to-ceiling window.

Tommy tried to wait, but he couldn't. He stepped to the edge of the door and peered in. One of the nurses was frowning at the doctor. The doc said, "I'm calling it. Two forty-six."

The nurse jotted something onto a piece of actual paper. Adjusting her cap, she said, "I really thought she was going to make it."

"Another week and she would have," he said. "We got one break: contacting next of kin will be easy, they're all right here."

"No partner or offspring?"

"None. She never married. One sister, deceased. Oh, and a brother in the service. In Brazil."

"What was her name again?"

The man in white opened an old-style clipboard—batteries were a luxury that the hospital reserved for more complex technology—and flipped through a couple pages. "Rose something. Ah, here it is. Rosalind Rhodes."

Chapter 13
Staying in Touch

"I remember a bad tooth I had once. The thing hurt like hell, and the dentist said it was beyond repair. Bad teeth run in my family." Gwen smiled as if to prove the point, although her teeth had been manicured to look just fine. "I hated the thought of going in for oral surgery, but I knew I had to, just to get the pain to go away. That's the way I feel now."

"Your mouth hurts?" Tony asked.

"My soul does. I'd be happy to go home right now, but I know we've got to go through some pain and get things done. It's for the best."

"Like a toothache. Pain first. I see, bella donna. But unless you have an idea for fixing the engines–"

Gwen cut in. "Isn't belladonna a poisonous plant?"

Tony was standing next to her foldout table, distracted by a small old-style book sitting at its edge. It was the booklet he had seen on Phobos. "It's also Italian for fair lady."

"Ah, I see. No, I have no ideas for fixing this old ship. But I do have some ideas for other options. Just need to send a message. A secure one."

"I would not send from the room."

"Right, I'm off to the hub. They've got secure lines there. Mi casa su casa, Tony. Feel free to stay or go. See you at dinner?"

"Until dinner," he said.

Gwen left him in her room and headed toward the center of the ship. She had been thinking about Europe's western seaboard, about the darkness and cold settling upon Brussels and Amsterdam, blanketing Bergen and Nantes and Lisbon, rolling south into Tunis, Casablanca, Freetown and Lagos. And with the waves of darkness came the inexorable tides of anarchy, bedlam born of failed infrastructure and government resources stretched too thin.

Perhaps it was time to get in touch with Melciéna Valentine. Though she had turned her back on humanity when it needed her most, the woman still had insights. She knew things. She was the magician. Even now.

The closer Gwen got to ship's center, the less gravity she could feel, until at last she was pulling herself, hand over hand, along the corridor railings.

As she approached the entry, she saw the mystery woman chatting with Captain Haegen. Her long blonde hair drifted around her head like a Renaissance halo. She spoke in hushed tones, but Gwen could hear.

"Did my message get transmitted?"

Haegen glanced just inside the door at a screen. "Yes, the queue is empty."

Gwen drifted to the threshold. Glancing with forced nonchalance at the nearest monitor, she spotted the last transmission's identifier: i6Ap. It looked familiar.

"Dr. Baré, welcome," the captain said. "What brings you here?"

The other woman floated by, drifting up the corridor without a word. Gwen raised an eyebrow.

Haegen leaned toward her. "She is a most enigmatic passenger. And speaking of enigmatic, how may I help you?"

"You have a secure line here, right?"

"I can set that up for you, certainly. Strap in at this station and give me the address, and you'll be on your way."

She did. In moments, the screen provided a readout of the transmission source—the cycling ship—and its destination: i7p. That was it. Almost identical to the destination of the mystery woman's message.

Gwen used the keyboard rather than the touchscreen or voice; she wanted privacy. She sat at the console, staring, thinking, hands hovering over the keys. She could say so many things. Where was Melciéna, really? What had she been doing all these years, and why get in touch now? What was her interest in Enceladus? She sounded concerned, even alarmed, that someone might be coming out to the abandoned ice moon. Why? Finally, Gwen began.

Melciéna:
After all this time, it would be wonderful to see you again, but we are on a cycler with a big problem. In fact, we will probably come sailing by your laboratories on Iapetus without a pause. The engines are out, so they can't tweak the path of the cruise ship to meet up with the shuttles. You seem to have concerns about Enceladus, as do I. Would it be possible for us to consult, perhaps when we're close enough for actual two-way radio conversation? And do you have any ideas for getting us off this tub? They've got about twenty passengers and a handful of crew. On a cruiser this size, the place is a ghost town. As for getting out of this, I think they're at a loss, and we don't have enough O2 or food to get all the way back. For now, guess we catch up on reading. Bye for now. Please advise of any options.
Gwen

She kept the tone businesslike, the content detached. She still wanted some space between the two of them. It was a long shot. What could a reclusive, eccentric old woman do about an emergency six hundred million miles away?

***

Tony stood at Gwen's table, all by himself, wondering what was hiding in the little booklet. Opening someone's journal wasn't like opening someone's email or underwear drawer, was it? He picked it up. Clearly, it was old. Creases scored the cover, and the pages were dog-eared. The paper was crafted of real wood pulp, made for sketching. But it wasn't a sketchbook. It was a diary. There were a few doodles, but a lot more words. He sat down and flipped through it, stopping on a section marked "Saturn, the fluffy king of rings." He began to read.

"Here I am, a full twelve years old and stuck in a place that is dark and wet and scary, with creepy monsters who walk up and down the hallway. And I will be a prisoner here for weeks. What did I do to deserve this? Oh well."

Chapter 14
Interval 4: Abandoning Enceladus, The Past: 2201

Sophia Baré's eyebrows made a little pyramid above her eyes, a formation that entertained Melciéna Valentine no end. But Valentine had to push her amusement aside to listen to the woman's tirade. Sophia's pilot, Reggie Barnes, stood at her side, stoic.

"Melciéna, you can't stay here. Not without your crew. This place will fall apart, just as all the others have. We're taking most of the nukes with us, and that will be it for both Enceladan complexes."

"The cities still have their bioreactors. Without all those people hanging around, they won't need nuclear power."

"Unless somebody taught those microbial colonies some new tricks, they won't run themselves. They'll collapse before the year is out."

"Sure, if no one tends them."

Barnes sounded urgent. "Dr. Valentine, please. You really can't stay here. Enceladus should have been deserted a long time ago. You seem to be the only one with a differing opinion."

Valentine shook her head, a faint smile softening her face. "Sadly, as with many things in life, we don't get to vote on what our friends and colleagues do. This is one of those occasions. But you make it sound like a death sentence."

"Well?" Sophia let the word hang in the silence.

Valentine leaned over and took both of Sophia's hands in hers. "There is something very special about this place, Sophie. You both know that." She glanced at the pilot and continued, "You know how much I love the Naiads, how much I've grown to love this place. I visit whenever my travels take me this way. There's something comforting in eternal darkness. Does that seem strange to you?"

Sophia pulled her hands away and peered through the bulbous window in the wall. "Like all the stickers say, embrace it. If it's darkness you want, you're about to get a whole lot of it. And there's nobody left out here to help you get out."

"Once the power goes, you're going to have a devil of a time getting up through the ice," Barnes added.

"Nonsense," Melciéna scoffed. "Everything has manual backups."

"Sure, but on manual, you've got to get into the airlock, slam the door behind you, and flood the compartment. Then there's no going back. Once it's flooded, there's no way to vent it again without power. So you go up for a couple klicks and then you're stuck out on the surface. There's even less infrastructure up on the ice than there is down here. What then?"

"Oh, Barnes, you're so practical," Valentine marveled. "How long has this woman put up with you?"

"Since before the accident. She's a patient woman."

"And he's a good pilot," Sophia hooked a thumb toward Barnes.

Turning back to Sophia, Valentine said, "Have you forgotten the joy of your childhood? This place..." She swept her hand toward the glass doorway and the corridor beyond. The ceiling towered above in a series of glistening Gothic arches. "This place is made of magic."

Sophia sighed. "Your own Ryugu. I know."

"Ryugu?" Barnes frowned. "The asteroid?"

"The fable," she said.

"Not familiar with the reference."

"Don't feel left out. I only know because Mel told Hugh to look it up once. That was a long time ago."

"Not so long. Do you remember?" Valentine asked.

"Japanese legend. Palace under the sea, owned by the dragon ocean god."

"Ryujin, yes," Valentine encouraged.

"And?" Barnes' tone betrayed impatience.

"You see," Valentine said, her voice tempering, "Ryugu was a crystal palace, a place where a day inside was equal to a century in the real world. And I've often thought that Thera and Atlantis West are that way. In the darkness, one loses all sense of time." She was staring off into some unseen distance, but she snapped back to the present. "The point is, there is history here that should not be forgotten. History and culture."

Two lumbering Naiads slogged their way up the corridor, not nearly as graceful as they were in the water. Their fish-dead eyes lacked the glint of intelligence. Did one of them nod to Valentine as they passed by? Were they even capable of such a thing?

Sophia leaned forward, her voice poison. "Tell me you're not staying for them, Mel. Not for those creatures."

Barnes spoke from the corner of his mouth, "They'll all be gone in a few months. It's sad, but they can't take care of themselves."

"Is that right, Melciéna?"

Valentine ignored the question. "My dear, it was an accident, and ruled so."

"That hybrid should have been destroyed, at the very least."

"I don't blame you for feeling so. That kind of pain stays fresh for a long time." A smile broke across her face. "But you two have lost your joie de vivre. Life should be something of wonder, of fevers and passions. The worlds are filled with magic, as long as you look. And sometimes it's a little bit of smoke and mirrors. Where is reality? Where is illusion?"

"Oh, I don't know," Sophia objected. "I'd say reestablishing the interplanetary network can be fairly magical."

Valentine stared at her for a moment. "I hope so. You'll be a lot happier if you really find it so."

"Don't get trapped down here, Mel."

As if to underscore her point, the lights in the corridor flickered. The farthest end of the passage went dark. Somewhere, an alarm chimed, like the tinkle of a music box.

Barnes fidgeted. "That's our cue, Melciéna. Time to go."

But Sophia tilted her head to one side, jutted her chin toward Valentine. "Smoke and mirrors, huh?" She glanced down the hallway, in the direction that the Naiads had traveled. "I know you're up to something, but I can't figure out what."

Barnes grabbed her by the elbow. "Well don't look so entertained about it. We're leaving."

"Goodbye, Melciéna," Sophia called over her shoulder, extricating herself from her pilot's gentle grasp.

"Farewell, you two nomads," Valentine said as another bank of lights died.

Chapter 15
Sisters

November 2229

Gwen entered her quarters just as Tony was tossing her journal back onto the table. She waved it off.

"It's fine, Tony, really. Feel free. It's not the most riveting reading; just a kid's log of travels from Mercury to Neptune. If you keep it on your bedside table tonight, you can use it as a sedative."

"I think it is very interesting, mi amica. At least the part of Saturn and…" His voice died out.

"Oh, you got to the Enceladus part, did you?"

Their eyes met. He studied her for a moment. "You stopped writing for a while, I think. Just what happened back there? To your sister? To your father?"

Gwen pressed her hands together as if she was praying. She closed her eyes and drew a long breath. "An accident. My father died."

"Your sister, she is okay?"

Gwen smiled. "Yes, Claire's fine." Truth be told, she had no idea how Claire was. But that was a discussion for another time.

Tony tilted his head to the side, curious. "Are you close?"

"Not as close as we used to be. Mom died in a decompression accident a few years after Dad died. It was a real one-two punch. We got pretty chummy for a while. Even before that—before Daddy died—we had our own language with scrambled words. Whenever one of us had a secret to share, they called out, *Gerthingmanchester*! And the other would reply, 'Secrets to share. Secrets to tell.'"

"Gerthingmanchester. Is it in England?"

Gwen shook her head. "Made up word. That was a long time ago. But diaries and journals eventually took the place of shared secrets, and we drifted apart. If anybody tells you that time heals all wounds, don't believe them."

"You have not been so forthcoming about Enceladus, mi amica. I thought you had only visited. This sounds like you were there for long time."

"A few weeks. Long enough. My father was helping to direct some Naiads who were repairing some kind of power equipment outside. It took days. My sister and I had the run of some of the corridors, but we found some places—secret lairs, we called them—where our parents surely didn't want us. One of those was an airlock in a closed section of Atlantis West. From its windows, we could watch the hybrids come and go outside, and we could see the ones that Dad was directing. They were very close by."

"He did not say anything to you about watching him work?"

"He couldn't see us. But somebody found us there one day, and he sent us to our rooms for an eternity. That's what it felt like, anyway. Once Mom and Dad sprang us to freedom again, we followed directions a bit better. But my sister wandered."

"She went back to your lair."

"Well, I wandered, too. Dad and them had been having some problems, and we liked watching the humans and Naiads working together outside. But I got bored and went back to the commons area." Gwen's voice sounded childlike, innocent, fearful. "I had forgotten my book, left it with Claire. Dad was in the commons. Something was going wrong out there—outside the habitats in the open water, where they were building additions. I could tell because suddenly there were grown-ups scampering here and there. They called him. I remember that when they called, the color drained from his face. It scared me. He said, 'Where's Claire? Where's Claire?' And I told him, even though I think he already knew, and he and I dashed down the corridor toward the observation dome, toward our lair. That's where the problem was, just outside, and that's where Claire sat, watching it all unfold." She laughed softly. "Good word for this: unfolded. Unhinged. Undone."

Gwen reached absently for the diary, held it in her hand. "When we got there, I yelled at Claire. I called her a bad girl, something like that. I wanted parental backup, but Dad was looking up at something. I remember that. And there was a screeching sound and a big crack in the glass where some piece of equipment had fallen against it, I think. Everything seemed to be falling apart outside. Dad grabbed Claire and he pushed us through the hatch." Gwen's thumb rubbed back and forth against the cover. Her eyes grew rheumy. "The water was squirting through the crack, just a misty little stream. We were on the other side of the door, the station side. Dad slipped on the wet floor and dropped my little journal. It slid through the door just as the safety protocols slammed the hatch shut. He was trying to get to his feet, but it was too late. We could see him on the other side, through a little widow in the hatch." Her voice was low, mumbling. "It was amazing how fast it filled with water. So fast."

"I am so sorry," Tony said.

"The damned hatches all have manual overrides so you can get people out, but I was a kid. I didn't know how to work them. Dad tried to show me, through the window, but there wasn't enough time and I didn't understand." Gwen paused, lost in thought. "Naiads. Hybrids. Mindless oafs. They watched through the glass. Didn't do anything. Didn't know how, I suppose. They played their own part in Father's death: sins of omission." Gwen looked up at him and handed him the book. "Here. Take good care of this. I need it back."

She was offering her colleague something of her past, some significant piece of her own history: the chronicle of the Baré universe. Tony's gentle tone showed that he knew it. "Of course. Thank you."

Tony left her to her own thoughts. As Gwen lay across her small bed, the monitor pinged. It was a message from Melciéna Valentine.

Gwen:

Discussing Enceladus over an open channel would not be advisable. I am sorry we cannot connect. Perhaps next time.

Perhaps next time? What did she think this was, a weekly knitting circle? There might not be a next time—for anything—if the cycler's passengers couldn't be saved soon. But it was the same old Melciéna: gone as soon as things got tough. Gwen felt the old anger return. It burned inside. And the eye was there, watching. She heard screaming again. This time, it was the voice of her father.

Chapter 16
Offering

"It's no good," Chevalier lamented. "I've checked private and corporate manifests, Jovian schedules, military resources, transports from Venus to Miranda. There's nothing available with the capacity to get out there in time. And our one fast transport at Iapetus just blew an engine. They're waiting on parts all the way from Mars."

"You'd think we would have enough stuff at our base on Iapetus to fix a damned engine." Carter Rhodes rubbed the back of his neck. "Mars. Shit. We've got the one hotrod vehicle that I could have dispatched from Ganymede, but it's been essentially out of range for a couple weeks, and the gap is getting worse."

"And the thing's only big enough for a couple people," Chevalier reminded him.

"It would have been enough for our needs."

"The press would have a field day with that: Earth military rescues own, abandons the rest. Martian media would be all over it."

Rhodes shook his head. "Like sharks to a bleeding seal. They do enjoy our little failures at times. Armand, why don't you ask Ben to send us up some lunch."

"Sir." Chevalier left.

Rhodes pondered the vast distances between planets, the inexorable dance of the spheres and how it continually confounded the miniscule plans of the human race. If those planets would only stay still. The waiting for launch opportunities sometimes taxed his patience.

The monitor pinged. Chevalier's face appeared. "Sir, message incoming from a Samuel Varner."

"Who the hell is Samuel Varner?"

"He represents Reginald Broadwalter."

"Broadwalter? The Martian billionaire?"

"That's him. Varner is stationed on the Moon, at the Copernicus complex, for close comms with Earth."

"Handy." Rhodes tapped his chin. "All right, put him through."

In moments, the stark face of Samuel Varner snowed into focus. He wore a dramatic, jet-black widow's peak atop a ghostly pale angular face, darkened with a five-o'clock shadow. His cropped hair bobbed in the low lunar gravity. "Colonel Rhodes, a pleasure," he said.

"Good afternoon. What can I do for you, Mr. Varner?" The niceties were out of the way. Down to business. He waited for the three-second delay between Sao Paulo and Luna.

"I'm contacting you on behalf of Reginald Broadwalter. Mr. Broadwalter wants me to let you know about an asset he has out at Iapetus."

They knew about the cycler. But how? Rhodes simply said, "Yes?"

"One of Mr. Broadwalter's yachts is outfitted with a combination of supercharged chemical engines and VASIMR ions. Sleek design for atmospheric flight. It's a burner, like your military grade transports. You take a cocktail of meds, strap in, and get plastered to your couch for a few hours as you blaze up to speed. Can be dangerous, but gets the job done. Mr. Broadwalter might be willing to make this particular asset available for the transport of your personnel currently aboard the cycler."

"That's very charitable of him. How many passengers can he take?"

"Two. Three at the most. But in the meantime, your resources could dispatch something to meet up with the cycler at a more convenient location, perhaps closer to Jupiter's orbit on the cycler's way back in."

Rhodes was warming to the idea. "That might work well. We have several such ships that than will be made ready soon. We were made to understand that the cycler's supplies might not hold out that long."

"Our yacht can ferry a few emergency supplies to tide them over."

"That sounds generous."

"Mr. Broadwalter is a generous man."

It was true. Broadwalter had made his name by establishing library hubs, hospital facilities in remote regions, even pumping up infrastructure in several critical areas on Earth in the darkest times. But Broadwalter was as much business tycoon as philanthropist. Rhodes was beginning to suspect the innocence of the offer. "What would your employer like in return?"

"He has advised me to tell you that this is a favor for which you owe him nothing…at this time. He may be in touch later."

"I would very much like to thank him directly, even if it's via a remote visual transmission."

"Mr. Broadwalter is somewhat, shall we say, reserved. He does not operate in that way."

Mr. Varner had made the understatement of the year, Rhodes thought. Broadwalter was a recluse, famous for dodging interviews and avoiding social engagements. Then again, Wentaway seemed to have produced a multitude of recluses, people who preferred to remain off the grid rather than deal with the various interplanetary crises. Varner continued.

"Rest assured that he is happy to contribute his formidable resources to this cause. We don't want anybody starving to death or asphyxiating on a distant cycling ship, now do we?" He dipped his chin slyly.

"No, certainly not." Rhodes had seen battle. He had flown rescue and exploratory missions to the most dangerous corners of the solar system. He had faced death and come out on top. Why then did he find this man so intimidating?

"I will relay your gratitude to Mr. Broadwalter at our next communications session, Colonel Rhodes."

"Mr. Varner, may I ask how you knew there was a crisis aboard the cycler? There's been no official notice."

He smiled. His face disappeared as he severed the connection.

Rhodes blinked at Chevalier. "These people drive me crazy. Can't we get a straight answer out of anybody these days?"

Chapter 17
Interval Five: Tommy

Thanks for the Memories

"I'm looking for something specific." The old man leaned against his cane, his face uncompromising.

"Hopefully, you've come to the right place," Tommy said. He was feeling insecure, staring at the clean-cut gentleman through his mop of hair. "The Madagascar Discord."

"Interesting word, isn't it? Discord. Sounds so civil; so urbane; so proper. I can tell you, it was none of those things."

"In 2194?"

The man nodded so exuberantly that Tommy thought he might fall over. "2194, that's right, son. Long time ago. Ancient history. But some of us were there."

"And it's not in any digital sources you can find?"

"Even had a military historybot try. No dice."

"Thank you for your service, by the way," Tommy said as he scrolled through files on his screen. "We haven't documented all our books in that section yet, and we're hoping to get more, but maybe something's come in. Shall we have a look?"

The two sauntered down a long, narrow aisle between high rows of shelving. The old man mumbled, "Just a couple place names I can't remember." He tapped his temple. "Growing old is not for the timid."

They turned through a doorway at the end. Tommy flicked on the lights. Next to the switch, a peeling sticker said, *Embrac th Darkn s.*

"This is the stuff that's still unshelved, undocumented. Some of it was just donated. But it's been organized, more or less. I'm thinking Madagascar should be in one of these two stacks if we've got anything." He could smell the paper and ink. As he pulled the tomes from the column, he felt the wood pulp of the covers. Some even had pseudoleather. The two men from two very different generations shuffled through the books, sharing the joy of the hunt. Finally, near the bottom of his own stack, Tommy spotted a book that might be what the patron needed. He pulled it out, gingerly.

"This is not in very good shape, even for a thirty-year-old book. Cover's about to fall apart, so be careful. The letters are worn off the front, but the title page says *Engagements in Southern Africa*. That look like it?"

The man sat in a nearby chair, the oversized volume on his lap. For a moment, he stroked the cover lightly. Then, pulling the cover open as if it was an Egyptian pharaoh's lost treasure, he began to leaf through the pages. He smiled, subtly at first, then with the abandon of pure delight. "It was a difficult time for me, but peopled with wonderful friends. Isn't it amazing? I marvel at how people wrote things down through the ages. On clay and stone and skins, parchment and papyrus. And paper. Beautiful paper. Ground up trees! Can you imagine? Yes, and they weren't just entering data into some distant server, but documenting, word by word, something important, something significant, with their *hands*. But then, then we got cocky. Put everything down in the minds of our computers. In the cloud and on hard drives. Of course, nobody bothered to think about electromagnetic pulses or rocks falling from the sky." He quieted. "Or a new dark age with candlelight and cold. Thank you, young man."

Tommy shrugged the comment off. "My job."

"Nonsense. You are doing something important here, my boy. Look around you. This place is magical."

Tommy left the patron to enjoy the old book. Across the lobby, he could hear an adult reading to some kids by the dimming window light. Not a digital voice; just somebody's mom. In that moment, he realized that he was not cut out to be an accountant like his dad or a programmer like his aunt—God rest her soul—wasn't made even for serving in the coastal or the interplanetary, like this old man peering into his old book. Like his Uncle Carter wanted him to be. He wanted to get his hands dirty in the ink of history, to resurrect the mummies and rebuild the ancient temples, to bring the past to life and preserve the memory of what had been lost. That was the moment when he knew.

Chapter 18
Top of the Loop

Saturn swelled quickly from a bright star to an elongated golden spot. To one side, Titan glowed as an orange pinpoint. The two passed by with each twenty-second rotation of the cycler, and the observation deck was the place to watch.

"We're getting close," Gwen said.

Tony followed her gaze through the portals. "You can almost see the rings now. And while we're watching out the window, I hope someone's keeping an eye on our oxygen and cuisine."

"It does beg the question. I wonder how much is left."

Gwen and Tony pressed their faces nearer the windows. Behind them, someone cleared her throat. They spun around to see the mysterious blonde woman.

"Oh, hello there," Tony said.

The woman offered a cool, humorless expression. "Would the two of you mind coming with me?"

Tony looked at Gwen. Gwen looked at Tony. She said, "Do you have any pressing appointments?"

"Not that I recall." He studied the woman.

Gwen turned to her. "Where to?"

"Follow me."

As the trio made their way down the corridor, a public announcement replaced the rare background music.

> Ladies and gentlemen, this is Captain Haegen. As you know, this ship usually picks up supplies as it meets shuttles and passes Iapetus. You also know that our close approach to Iapetus has been cancelled for this cycle. I want to reassure everyone that we will have plenty of supplies for our return trip. We will resupply several times in the near future. Thank you for your attention.

"What, did they find a deli out here?" Tony asked. "Where are they gonna get food?"

"Something to breathe might be nice, too," Gwen said.

"Oxygen is useless unless it is used for digestion. The richness of life."

"The ship doesn't need oxygen, per se," the mystery woman called over her shoulder. "What it really needs is carbon dioxide filters. Those don't take up much room." She turned away again, and continued to lead them in silence. Soon, it became evident that she was heading for the airlock section. Her hair began to drift, fanning out as they approached the weightless center of the ship. Gwen's offered a complement in brown. Tony's kinked mop stayed put. They were using the handrails as they reached the first hatch. She opened it and gestured them to follow her inside. Gwen glanced at Tony, who looked dubious.

"Don't be shy," their guide encouraged. "I thought we could use some privacy." She sealed the hatch behind them. "Please allow me to introduce myself. My name is Taina Maes. I know who you are."

"Have you been babysitting us?" Gwen said with mock appreciation.

"I was sent to be your reserve option, actually. In case you needed one, that is."

Gwen's eyes narrowed. "Sent? By whom?"

"Someone who is interested in your wellbeing."

"And does this someone have a name?"

"Reginald Broadwalter."

Tony's eyebrows rose. "The magnesium magnate? Boyhood billionaire?"

"The very one." For the first time, Taina offered a smile.

But Gwen had more questions. "So how do you fit in? And how do we fit into Broadwalter's plans? Should we assume he has a vision for getting us out of here? Is that what this is all about?"

"Yes, partially. All you need to know for now is that you should be ready to leave at a moment's notice." There was a roguishness in the way she said it, a playful tenor Gwen had not heard before.

"Is the good Signore Broadwalter camped out here somewhere?" Tony asked. "Iapetus? Titan?"

"He would rather not say, I'm sure. But he wanted me to tell you before anything happens."

"Anything like what?" Gwen pressed.

Taina cocked her head and studied Gwen. "You are a curious type, aren't you? I will be in touch, but please don't discuss this with our fellow passengers." She turned, unsealed the lock, and drifted quietly out the door, back toward the passenger accommodations.

Tony and Gwen stared at each other for a moment, and broke out into gales of laughter.

"Yes, mi amica, this is the strangest voyage I have ever taken."

Gwen started back toward the rooms. "I'm going to do some digging."

"I'm going to do some snacking. See you at dinner."

"How do you keep your girlish figure?" Gwen chided.

Tony slapped his abs. "Good metabolism. Runs in family."

***

Tired of the pseudo-cosmopolitan cuisine of the *International Lounge*, Gwen and Tony settled in at *Millie's Southern Cooking*. Gwen confided, "I've eaten in the

French quarter of New Orleans. I suspect Millie is from some place like Bozemann or the San Fernando Valley."

"Southern cooking in Italy is quite different. I have no comparison. And what did you find of our cryptic guide?"

"Cryptic! Look at you and the English language. Good job." Gwen popped a couple sunflower seeds into her mouth, carefully.

"Those seeds," Tony said. "Like gold to you."

"I'm rationing myself. Who knows where the next stash of them might be? So for her name, maybe nothing is nothing. But it does have interesting roots. I did a search. Taina is Russian for 'mystery.'"

Tony smiled, leaning forward. "Really? But then she doesn't look very Russian to me."

"Ah, that's not all. It's also got a Scandinavian meaning: 'She knows.'"

"That is rich, as you would say."

"I would bet her real name is nothing like Taina. But it's entertaining."

Dinner arrived with much conjecture about the mystery woman. As Gwen washed down a remarkably well-crafted beignet with chicory coffee, an announcement came over the ship's system.

"Passenger Taina Maes to the bridge, please. Taina Maes to the bridge."

At that moment, their personal devices paged them to do the same.

Gwen frowned. "Why didn't we get an impressive announcement over the audio system? A text is so mundane. I feel slighted."

"Maybe *Taina* doesn't believe in personal electronics," he said.

"Or maybe she wanted us to know she would be there, too. She wanted us to hear."

Tony shrugged, noncommittal. They made their way toward the ship's hub. The Captain greeted them at the entry.

"Welcome! Please, come." She swept her hand toward the control room.

Taina floated at a central screen with a crewmember. The crewman said, "There it comes. Right... there."

"Yes, good trajectory," Taina said. "Good velocity. Looks nominal." She turned to Gwen and Tony. "I wanted you both to be informed of the arrival of our little ship."

Gwen started. "Ship? Here?"

"It is small. Too small to solve all the problems of our cycler. But it's carrying supplies to tide everyone over a little. And it's fast enough to get you where you need to go."

Tony whistled. "That's a beautiful little ship."

"Yes. Personal yacht of Reginald Broadwalter. It's called the *Belvedere One*."

"Means in Italian 'beautiful view,'" Tony put in. "Except for the 'one' part."

"Is the good Mr. Broadwalter driving?" Gwen asked.

"No one is at the moment. Automated docking sequence. But eventually, I will be. I'm your pilot."

Mystery woman, indeed.

Chapter 19
Planetary Yachting

To the crew of the Mars/Saturn cycler, it felt like an early Christmas. Reginald Broadwalter's small yacht was packed floor to ceiling, with fresh fruit, packaged and condensed foods of every description, and the all-important carbon dioxide scrubber filters. Even the cockpit had been stuffed full of supplies.

With the *Belvedere One* offloaded, Taina invited Gwen and Tony aboard. Plush velour padded the interior of the airlock. Wood paneling formed sweeping lines of rich umber running the full length of the ship's interior. The front bulkhead, the wall that must have backed the cockpit, displayed soothing landscapes of Earth and Mars on several large screens. Gwen floated to the front and peered through the hatch. In the windows, Saturn glowed like a child's toy, still a long way off. The cycler would come no closer, but the yacht would.

The yacht's captain spoke little, and volunteered even less. All they knew was that Broadwalter's yacht would take them exactly where they needed to go on Enceladus. But Gwen wanted details.

"Like how much time do we have at Thera, and when will you pick us up? And how will we fit more than one nuclear plant aboard this little cruiser?"

Taina's lips twitched into a subtle smile. "So many questions, Dr. Baré. My boss warned me."

Gwen took notice. "Your boss warned you about what?"

Suddenly, Taina looked as if she had been caught rifling through someone's purse. She tried changing the subject. "We will enter Enceladus orbit at 09:00 tomorrow morning. There are hammocks in the back, along the sides behind the lockers where you stowed your gear. Captain's quarters are at the rear. Landing at Thera should be at about lunch time."

"What was it that your boss warned you about? Does he think he knows me from somewhere?" Gwen demanded. "Have we met?"

Taina flushed. "Dr. Baré, you do not know as much as you think you do, and life will not always give you the answers you want. Now, I must tend to the ship. Feel free to dispense coffee or tea from the unit at the back of the main compartment."

They had been dismissed.

As Tony filled a zero-g cup with espresso, Gwen squirmed. "I could swear I've never met the guy, but did you see how Taina was avoiding the question? She obviously didn't mean to, but she clearly implied that her boss told her something about me. How would he know?"

Tony handed her the cup, but didn't take his eyes off her. "Gwen, this is no time to look the horse present in the gums. We must do nothing to endanger the generousness of Mr. Broadwalter, va bene?"

"Generosity. I suppose you're right. I just wonder what's going on."

He sounded like a stern father. "What is going on is Reginald Broadwalter, king of the asteroid mining, is bailing us out in style. That's what I think is going on. Please relax. We are almost to our mission, despite big odds."

Gwen took a deep breath. "Yes, we are," she said, stuffing a fistful of sunflower seeds into her mouth, chewing with far too much gusto.

***

The incoming message from Earth appeared on Gwen's monitor right on time. Carter Rhodes was finally back in contact, and the fact gave Gwen a sense of relief.

TO: Baré, Gwendolyn
Vincenzi, Antonio
FROM: Rhodes, Carter
RE: Enceladus
TIMESTAMP: Nov27/2229/8:48:07GMT

Drs. Baré and Vincenzi: By now you have been debriefed by Captain Maes. Thanks to the generosity of Reginald Broadwalter, your mission is back on track. Current plans call for two cargo pods to be deployed from Titan. The first will be supplied by Mr. Broadwalter, and arrives tonight at the Thera site. This one is smaller, but I'm sure it will afford enough space for two nukes and at least a couple bioreactors. You can pack it up and leave it for pickup, and then you'll be shuttled by Broadwalter's yacht to Atlantis West. The second pod is one of ours, a military grade pallet that can take anything you can get your hands on in Atlantis West. The yacht will return to Broadwalter by then, but our military transport should arrive shortly to take you two and your precious cargo home. I really appreciate the hard work you are both doing. Rhodes out.

***

The *Belvedere One's* bridge was a tight fit for the trio. Like the rest of the ship, it dripped with opulence. Where the engineers had outfitted the functional cycler with padding and metal, wood framed every one of the yacht's windowpanes and control panels. Gold piping accentuated edges and corners. The central light fixture was more Tiffany than utilitarian. Gwen and Tony stationed themselves in seats directly behind Taina. Heading for Enceladus orbit felt like the opening of the Indianapolis 500.

"This thing's amazing," Tony marveled.

"Some day we'll be able to go a whole lot faster. Some day soon."

"You're a fan of science fiction?" Gwen mused.

Taina's gaze remained focused on the console. "I'm a fan of advanced engineering."

A golden glow bathed the ceiling and portside wall of the chamber, emanating from Saturn to their left. Ahead, a plain of blue light spread before them, the misty E-ring that marked the orbit of their destination. From the right, the nearly silhouetted orb of Enceladus came into view, crowned by a thin crescent of glowing ice. As the moon slid along the E-ring, it swelled into a strange and wondrous world of cracks, canyons, and cratered plains. Gorges rifted the glowing white provinces with deep blue hues. Somewhere to the south rose the plumes.

"I'm taking us from north to south," Taina explained, "down along the night hemisphere. We'll fly under the moon and head back toward the equator on the day side."

Gwen watched Taina working. In her element, their mystery woman was no doubt capable, even exulting.

The Sun dropped swiftly behind the northern limb of the iceball. The edge of the moon glowed in blinding light, but the really interesting stuff lay at the terminator, that twilight edge between night and day. There, every detail stood out, the mountains and crags rising from darkness into the low sunlight, a Renaissance artist's chiaroscuro. The blackness of night slithered through deep canyons into the daylit side, while crater rims inscribed semicircles in the dusk. Enceladus' crescent thinned to a ragged line, then finally disappeared, gobbled up by the night side of the moon. The dark landscape below smoldered in the backlight of Saturn's amber clouds. The dim terrain reminded Gwen of the cracked surface of raw dough. The crevasses cut through crater rims and encircled rises in the topography. In minutes, light flooded the cockpit from below. Sunrise at the south pole of Enceladus was a spectacular affair. The curve of the moon burned a brilliant white arc into the sky before them.

The surface here gave the appearance of frosted bathroom glass. Rainbows of color scattered across the ice. As the ship approached the famed volcanic Tiger Stripe region, the polished ice gave way to powdery snow. The soft grey banked up in hummocks and gentle ridges. Damascus Sulcus glided into view below them. A parade of gauzy streamers drifted above the great aquamarine depression, melding into a great plume. As the Sun winked in and out behind the jets, a spectral halo followed it. The curtains of vapor undulated against the stars, geysers hundreds of kilometers tall, spewing briny jets from the deep waters below. The ridges bracketing the Tiger Stripes rose up against the black sky, breaking the graceful arc of the horizon. Only a handful of very young craters punched into the rolling wilderness. The ice near the geysers became a tortured series of crests and twisted ranges, a wasteland sculpted by the forces of cryovolcanism.

The yacht made its way past the geyser-cleaved valleys, on into the folded plains in the crook of the Labtayt Sulci's "Y" further north. The canyon, with the top of its Y facing south, branched into Cashmere Sulci, which spread east and west. Hues of ultramarine and cobalt lingered in the bottom of the abyss where the ice was densest. Taina banked the ship to starboard and headed for her landing target on the rippled surface below. "Not the smoothest spot for a landing," she whispered, her eyes locked on the approaching wilderness.

Somewhere beneath the landing pads, they knew, spread the underwater settlement of Thera.

Taina called out the numbers as they descended. "Altitude thirteen. Twenty downrange."

"Miles or kilometers?" Gwen asked.

"Kilometers," Taina said, peering through the expansive front windows. The horizon tilted to the left, then straightened. Ragged mountain ranges drew serpentine ramparts around and through craters, crossing hollows and breaching crater walls as if they weren't there. The uplifted ice showed yellow and purple shades beneath, brines tinting the matrix, but as the piers and blades rose into the sky, they became sun-darkened, with hues of red and brown where the sunlight cooked methane and ammonia into dark tholins. Some of the rearing promontories glowed from within, the warm rays filtering through the translucent ice.

"We should see it any moment," Tony said, straining to glimpse details below. Despite the great distance to the sun, the brilliance of the ice blinded them to the smallest features.

Tony pointed. "There!"

Gwen followed his gesture. She spotted it. The complex was small, with two circular landing pads—black against their snowy surroundings—a small dome between, and a comms station atop a rise off to one side, bristling with antennae. Taina piloted the *Belvedere One* with skill, dropping gracefully onto the painted crosshairs of one of the pads.

"Nice job," Gwen said.

"Before they shut down the power here, they could bring ships in on auto. But I don't mind flying in myself."

"Spoken like a true pilot," Gwen said.

"I'll just park here, for now."

Tony sounded animated. "Si, keep the engine running."

"Don't worry; I'll be listening in case you guys need anything. The cargo transport pod will be landing late tonight. Once you've loaded it up, I'll ferry you to Atlantis West. When you're done there, you'll hitch a ride—you and your power plants—on a new military cruiser, courtesy Colonel Rhodes. The thing's almost as fast as the *Belvedere One*, though not nearly as attractive. Big enough to take our two large cargo pods, though. So, suit up, get your gear, and I'll send you on your way."

Gwen had listened to Taina's scenario carefully. It all checked out. At least Broadwalter's pilot and Carter Rhodes were on the same page. After all the false starts and long waits, the frustrating cancellations and miraculous solutions, the time had come. Gwen and Antonio Vincenzi were now fully committed to their visit down to an alien seabed.

***

Gwen could feel the pulse in her temples. It hurt to move her neck; every muscle ached with tension. She hadn't been out in a full-vacuum pressure suit in ages, and it gave her the creeps. The pungent smell of sweat permeated her suit's air. The acrid taste of salt and burned ozone singed her tongue and throat, the essence of pure, raw fear. She was no professional cryonaut. What was she doing out here?

From the Thera complex viewpoint, Saturn loomed low over the northwestern horizon, a banded behemoth filling a third of the sky. Its rings stood at a dramatic angle. Seen nearly edge-on, the rings appeared as a thickened straight line, but the shadows they cast across the globe of the planet were spectacular.

Gwen and Tony hiked across the frosted ground toward the central dome. The heart of the complex was far smaller than any of the McMurdo stations, just a humble outpost with access to the seas below. But that's what made it so important.

The ground squeaked beneath their boots, and more than once, Gwen lost her footing as her feet slipped on the polished ice hidden beneath a thin layer of powder. The ice reared out of the plain in fins and buttresses, glistening like crystal sculptures. It bore the same hues as the dense ice in a glacier, ranging from turquoise to aquamarine. From here, the travelers could just make out the geysers rising in the distance, somewhere beyond the southern horizon. The plumes rose and fell in a cadence, the heartbeat of Enceladus.

High above and farther north, the distant orb of Titan stared down with its foggy orange countenance. Beyond it, starlike from here, floated Iapetus. Gwen wondered: was Melciéna up there, so close by? Why would she not come to help, when a billionaire recluse would? It was puzzling, frustrating.

The base dome was a dull gray, punched through here and there by portholes. The interior looked completely dark, dead, uninviting. This far from the sun, everything was cold, but the interior presented a chill deeper than temperature alone could give.

"See this?" Tony said as they reached the entry. "This used to have pretty little lights indicating what goes on inside. No power now." He scraped ice from the keypad. "Looks like long times since this place was alive. Hope there's still pressure. I'm plugging in."

Gwen fought back a smile. The subtleties of language could be easily lost. Tony pulled out a portable power pack and affixed it to the dead lock.

"There are your pretty little lights," she said as the unit powered up.

"Welcome, my lovelies. Tell me, is there air back there, or do you have another nasty surprise for us? Like water from below?"

"Is that even possible?" Gwen asked, alarmed.

She could just make out Tony's shoulder shrug inside his bulky suit. "Doubtful. At certain times in the orbit, the stresses allow the plumes to jet up and out from subsurface reservoirs, or even from the ocean itself. It makes its way through the weakest fractures in the crust. Very opportunistically, yes? That's farther south, of course. But here, we have made our own fractures, conduits through the ice. These weak spots, they might be overcome sometimes and water comes all the way up. But never happened in all times that people were here. The crust is thicker here than in the Tiger Stripes. Not as pretty, but more firm."

"That's reassuring."

"Yes, good, mi amica. We don't want to be nervous here. Must be sharp. Pay attention. Who knows what is waiting?"

The small screen in the door's console flashed to life, streaming lines of data. "Ah, yes, here we are. Pressure is good inside. No light, but we can fix. Everybody's ready?"

"This 'everybody' is," Gwen said, breathing deeply.

Tony punched a command and pulled a heavy lever next to the keypad. They could see air venting into the vacuum of space, feel it in their feet, but they heard only the quiet fans of their environment suits.

"Why is the air venting out instead of recycling?" Gwen asked.

"It beat me. Something set wrong, maybe a stuck valve. The place is old. Not used for long times."

The panel indicated hard vacuum inside the lock. They swung the hatch open and stepped in. The spacious chamber was designed for loading large cargoes, with a no-frills corrugated metal floor and ribbed walls with tie-downs. Its interior was slick, bathed in a thin sheen of fresh ice. "Must have been humid inside before we opened the lock," Gwen guessed.

Tony looked her in the eyes through their helmet visors. "Are we ready?"

She gave him a thumbs up. He slammed the hatch shut. The only light now came from their small helmet lamps and the sunlight outside, streaming through a tiny window in the hatch. Tony plugged in his power unit again. "Here goes niente."

The interior controls flickered, darkened, and then blinked back on. Gwen saw Tony give her a grin, but concern played across his eyes. He tentatively pushed a button and air flooded the compartment. It remained dark, but the console indicated full pressure. Tony reached for his collar, but Gwen put her hand on his arm. "Let's wait until we see the conditions inside."

"Yes, yes. Of course."

Gwen grabbed the handle, hesitated, then twisted. It was difficult. She grabbed with the other hand and it gave slightly. Tony reached over. The two of them wrestled the hatch open. Fog billowed into the hallway from the frigid airlock, scattering quickly. Gwen checked the monitors on her wrist. "Pressure's good. CO_2 and oxygen levels okay. It's cold in there, but something's been warming things up a bit, enough for us to be okay. Shall we?"

They pulled their helmets off. Vapor poured from inside their suits, then faded. Gwen shook her hair out in the stale air. Tony pulled his ponytail from inside his collar. "That's better."

"Smells like a paper factory with a sewer backup."

Tony sniffed and grimaced. "Yes. Moldy. Mildey?"

"Mildewy, I think is what we're after. If it's not a word, it should be." Gwen looked around. "That lock is plenty big to handle a reactor core and coolant system. At least the models I think they've got here should fit. And this corridor should be adequate. Now we just have to see what kind of elevator they've got."

"As I remember, the lift is every bit as large as the lock."

"Good. Let's go see."

19 Planetary Yachting

They headed down a gradual incline, far enough to know they were now well below the surface.

***

Armand Chevalier rubbed his hands together as if warming himself by a campfire. "They're safely down. Safely there. Good, good."

Rhodes scowled. "There is nothing safe about being on Enceladus, Armand. Not by a long shot. Not anymore."

"Well, of course not. But no missiles. No hostile action. That counts for a lot in my book."

"I don't like relying on mystery billionaires and luck. Not a good combination."

"All we need is for our luck to hold out just a little longer."

Rhodes stood and walked over to the window. Night had fallen on Sao Paulo. The glittering of the lights a mile below mirrored the swarms of airborne vehicles above. Life was returning to the big city.

Rhodes shoved his fists into the small of his back. "While we're rubbing our rabbits' feet and picking four-leafed clovers, I'd like to stack the deck. Stack it in our favor for a change."

"How would you propose to do that?"

"It's a variation on our Plan B. Instead of a large military operation at a long distance, why not take it a step down? We can still prepare for any eventuality that might involve a large force, but in the meantime, instead of reacting, we get proactive. Send a small strike force out. Station them nearby. Say, in orbit at Titan or Iapetus."

"Iapetus is far less busy," Armand said. "In fact, we have a frigate in orbit already, and troops on the surface who can transfer whenever you want."

"Yes, good. We have that ship standing by in Iapetus orbit—quietly, mind you—and then if we need to project force, we've at least got a forward position established. But we don't wait."

"No harm in that, I suppose," Armand mused. A stricken look crossed his face. He spoke up, "Very good strategy, sir!"

Rhodes turned on him. "I've had it with your insubordination! Go stand in the corner."

For a moment, Chevalier exhibited the deepest shock. A slow smile broke across Rhodes' mouth, and both men laughed.

"We get far too serious, Armand. Shall we get out of here for the day?"

"Yeah, sure. Sir." He grinned.

"Dismissed, soldier," the Colonel exclaimed. Chevalier saluted. Rhodes returned it.

Once Chevalier closed the door behind him, Rhodes paged his secretary. "Any news from Enceladus yet?"

"Nothing yet, sir."

"Any communication from Broadwalter or the yacht?"

"Not at this time, sir."

"Keep me posted of any progress reports. I'll keep my personal comms open all night."

"Will do, sir. Good night."

They were down there, on the ice, inside the Thera access point by now. What had they found? Darkness? Ruin? Damn these planetary distances. They made it far too easy to ignore communication protocols, and civilians were the worst at staying in touch. But on the other hand, if the missile-happy thugs out there were listening in, a little silence might be just the ticket.

He was too impatient to wait. He went home and watched old holovideos of the world before Wentaway. The stories were undoubtedly romanticized, but his life could use a little romance just now. Tomorrow would bring its own troubles. Probably big ones.

Chapter 20
The Water and the Wattage

With their helmets off, they could hear a popping sound and smell the aroma of a summer lightning storm.

"It's coming from down there," Gwen said, gesturing at the dark corridor that neither of them really wanted to go down. A flash of purple light accompanied another pop.

Tony craned his neck. "Yep, something's arcing. Shorting out."

"Yes, but here's the power distribution unit." Gwen opened the access port. Inside, several breakers glowed red, a record that they had flipped to counteract a short or electrical fire. "After all this time, I expected these to all be red. Or dead. At least there's power coming from somewhere."

She plugged in a handheld and read the circuits. "Looks like this one's the lights for this level. We'll need light when we bring all that equipment up."

"If there's anything left to bring."

"My, aren't we cheery?"

Tony jabbed at one of the glowing buttons.

"No!" Gwen barked. "Don't touch. You can get a shock if the place has flooded. Might be water we can't see. The airlock should have been dry, but it was covered in ice. Probably the humidity in the air."

She prodded at the keyboard with several tools. Gingerly, she tapped the tab. Tony peered over her shoulder, and observed, "Niente. Nothing. Do you enjoy when I make obvious observations?"

"Very much so," she said, tapping at another. "If we want light, we gotta fix something, and I bet it's down there with all those fireworks. Let's just hope it's dry."

They made their way into the darkness, toward the shower of sparks sporadically lighting the far end of the walkway. As they neared, they could hear water. The smell of wet concrete mingling with the taste of ozone reminded Gwen of a thunderstorm in Colorado.

"Maybe it's a slow trickle," Tony said hopefully. But the closer they got, the more noise they heard.

"Maybe whatever little river we hear is flowing away from the place we need to be?" Gwen suggested. But the closer they came, the louder the water thundered. Tony gave her a withering look.

They reached the entry hatch just as another shower of sparks flew from inside, glowing bees swarming the air. As she cleared the edge of the doorway, Gwen saw it. On the far wall, a row of electrical assemblies disappeared behind a bona fide waterfall gushing from a crack near the ceiling. Several side streams jetted from smaller fissures, combining forces with the main tributary. The surge flowed across the floor into a web of cracks where a drain used to be, disappearing into the lower levels.

"How far down are we?" Gwen asked.

"Not far enough," Tony said, anticipating her question. He sniffed. "It's not sea water. The ocean is a long way below. This is coming from something near the surface, maybe even a storage tank within the station."

"Well, I doubt that. A storage tank wouldn't last long. Look at the wires."

Gwen was right. The cables draped through the stream of falling water had all but rotted away. Where the insulation was gone, the wires were eroded through, in some places entirely severed.

"Maybe a subsurface aquifer, big lake inside the ice crust. Lots of damage, but what needs fixing?" Tony wondered aloud.

Gwen pulled up schematics, studied the little diagrams, and gazed at the wall. "Not too bad. I can take wire from those other cables there, which lead to environmental control for some of the lower levels—no one is there to breathe anyway—and splice them over here so we can enable that breaker again. We just need wire with full insulation so Niagara can't get to its insides."

"What can I do?"

Gwen thought for a moment. "Maybe see if the elevator works, but don't go down by yourself."

"Don't worry. I have automatic cutoff switch for scary places."

"Wish I did," she muttered, heading for the thundering cascade.

***

Unlike the airlock above, the lights on the elevator were all aglow. They were the only light in the portico. Though he didn't think it was possible, the hallway here seemed even gloomier than the corridor with the sparking waterfall. Tony entered the lift cautiously. He didn't want to drop to the bottom of the sea until he was good and ready.

The elevator was a sealed cabin, airtight and designed to travel through both air and water. It rocked slightly as he stepped in. A flood of memories assaulted him, of a bright elevator, bright corridors, a brighter time. The chamber was elliptical with curved walls and a flat floor and ceiling. Tire marks, scrapes and gouges painted calligraphy across the metallic floor. Tony studied the control panel. He didn't remember its details from his visit long ago, but it seemed fairly straightforward: a small window, darkened now, spread across the top of the panel to call out levels as the cab made its way down into the depths. Below it were a series of boxes, illuminated on an ancient screen. Some were faded almost beyond recognition, while others bore familiar labels like "Surface" and "Thera Level Four." Still others offered more mysterious labels: "STEVII," "Biconic," "Engage," and "Abort." An illuminated pad at the base bore the inscription "Command," and below it, a keypad with numbers. Tony tapped the Command bar. The dark window at the top flickered. A vertical line flashed at the left end, much like the cursor on old-style computers.

"Interessante," he muttered. He tapped a random number on the keypad. The number two appeared on the screen next to the flashing line. "Okay, she works," he said. That was enough for now. He turned to step from the cab when his environment pack inadvertently brushed against the pad. A chime rang out, the door slammed shut, and the screen filled with the words "Thera Level Two." For the first time in decades, the cab lurched downward into the abyss.

Chapter 21
Drop

"No!" Tony hollered. He was on the floor, knocked off his feet by the sudden acceleration of the car. Was it in free fall? In the low Enceladan gravity, its drop would be gradual. But the cab careened downward in some sort of powered descent. He steadied himself against the wall, finally standing by the keypad. What to do? The elevator was descending at breakneck speed, and he could feel the pressure increasing. He sealed his helmet, activated his suit, and frantically studied the panel. The top readout now had a new display. It read, "Vertical distance to requested destination: 21.8 km." The ice crust here was just a couple of kilometers thick. That meant the cab was headed toward a 20-kilometer-deep stretch of ocean. The distance was counting down. So many buttons and labels. His eye finally settled on one: ABORT.

His finger hovered over it for a moment. What if it meant something else, something he wanted to do even less than a solo trip to the ocean floor? He tapped the key. Nothing happened. He tapped it a little harder. The whine of the descending compartment abruptly decreased.

The elevator came to a full stop. The lights flickered. Darkness. Silence. The elevator was frozen in place. Then, the lights sputtered on again and he could feel the upward acceleration. With a sigh of relief, Tony watched the door slide open. Gwen was on the other side. She was asking something. He pulled his helmet off.

"What happened? I thought you were going to wait for me."

"Elevator has ideas of its own. Luckily, it didn't go long. It looks like she's fully operational. Oh mio, you're a mess!".

Gwen's mahogany hair was plastered flat against her scalp, looking like a pile of chocolate pudding. Water dribbled from her suit, splashing around her feet. "I had to get a little wet. I've repaired the main cable in Enceladus' Trevi Fountains. When we throw the correct breaker, we should have lights."

They made their way back toward the entrance. The electrical junction now displayed several yellow glowing tabs, including the one marked Central Illumination. Gwen reached over slowly. "Here goes nothing." She tapped the button. It turned from yellow to green. The corridor flooded with light.

Tony clapped. "Looks like we're in business. Good job. Molto bene!"

"More than that, if I fixed what I tried to fix. Come over here." She ushered him into the next room and tapped a touchscreen. "Yes, you see? Now we have readouts from down below, too."

"You killed two flies with one rock."

She looked as if she was about to correct him, but instead she smiled. "Let's see what's going on down there."

The two studied the screen. The system charted temperature, pressure and other variables from the different levels of Thera.

Gwen pointed at the screen. "Temperature drops down at level five, all of a sudden. And the pressure goes up. The changes are abrupt. Flooded?"

"Makes sense. We may need to work underwater after all."

"Not necessarily. My blueprints show the nuclear reactors at higher levels."

"Let us hope."

"And I've shunted power into the lower levels for light." She sounded proud.

Tony didn't hold that against her. She had done a mighty deed. He said, "Excellent; easier than working with helmet lights."

"Much. Should we…go down?"

Tony reached for his helmet again. "Cannot put it off any longer."

As they headed for the lift, Tony said, "I must warn you, it is wild ride time on this thing. My head almost hit the ceiling."

Gwen tapped her visor. "Good thing we brought our crash dummy outfits."

The reference puzzled him, but there were more important things to attend to than linguistic nuances. "And your diagram tells us to go where?"

She aped a pretentious accent. "Level six, please."

"Hold on." He sealed his helmet, pumped the suit, and punched the elevator's button.

***

Even at the lift's frenzied speed, the ride to the seafloor would take them a quarter of an hour. They stood for a moment, facing forward, not looking at each other. Gwen began to laugh.

"Che cosa? What?"

"Look at us. We're riding this thing like it's a corporate lift full of strangers. All that's missing is the elevator music."

"It is one of my habits," Tony grinned.

"Did you ever think you would be doing this? Taking an elevator to the ocean floor of a Saturnian moon?"

"More than one time, for me," he joked. "No, life is full of surprises. My parents wanted me in the family business, and my Uncle Marco? He had own designs for me."

Gwen's curiosity took her. "Like what?"

"Nutrition. Did you ever hear of Toxic Waist diet?"

Gwen adopted the voice of an announcer. "'Toxic Waist: figure control for young and old.' That one?"

"That is the one. But our 'cleansing flushes' and 'detox smoothies' never interested. Not to me. Personally or professionally." He grimaced. "I always loved living things, botany and zoology. But I like what I have ended up doing with glaciers, too, so here I am."

He went quiet. In the silence Gwen could tell he was dreaming of a life that never was, a life that almost had been. Tony shook himself back into the moment. "Did you end up where you thought?"

Gwen tried not to think of the speeding elevator, of the kilometers of frigid waters now accumulating above them, with so many more to go. She controlled her breathing. "I've always loved the ebb and flow of energy, whether it's in biology or electrical conduits. But I never expected to see something like your Trevi Fountains in an abandoned access to a deserted submarine city. There were times, especially when I was just starting out professionally, when I thought about other paths I might have taken."

"Like?"

"Not the standard. Not a doctor or a corporate president or a physicist. I always liked the idea of exploration. But I was born too soon after Wentaway. I don't think anybody will be exploring for science's sake for some time. Or maybe I came to the scene too late. I would have loved being on the early expeditions to the Galileans, or on the Stern group to the Pluto/Charon system."

Tony frowned. "Too cold. Too slow. It took years to get anywhere back then."

"True enough. But new worlds, new frontiers! Seeing wonders for the first time in human history? That would be something."

They both quieted, lost in their own thoughts, as the numbers grew above and shrank below. The elevator traveled within a clear vertical shaft, and through its small window they could now see muted light filtering up from below. Outside, the elevator's running lights illuminated dust-like clouds floating in the seawater, the apex of the Enceladan food chain. The life within these primordial oceans resided in the microbial realm, a Lilliputian kingdom. But it was life, nevertheless. Life on its own terms, separate from the terrestrial Earth/Mars life chain, independent of any other world, intrinsic to this very distant, very cold worldlet.

"And what does your sister think of all your travels? Is she like you?"

"Claire and I are polar opposites. She's become quite the homebody. Really smart, but she lost all of the wanderlust. Quantum systems analyst."

"Does she have a family?"

Gwen stiffened. What was with the third degree? Then again, Tony was just interested, and she knew she was being defensive. But did she really have to go into it? Here? Now?

"Claire's married to a wonderful man who I barely know. She has a son who—again—I barely know."

"Are they far away?"

"Earth. Philadelphia."

"So close to the interplanetary hub on the east coast. Close to Venus. You don't get to visit much?"

"We…aren't that close." And why is that? Tony could have asked. He was probably thinking it. But what would she even say? *Claire has a lot of baggage. She probably never wants to see me again. She holds me responsible for our dad's death.*

Instead, Tony said, "It is so hard being far from famiglia, yes?"

"Yes," Gwen echoed robotically. "It is hard."

The elevator began to slow. The monitor read "1.8 km to destination." The cab was on its final approach to the infamous Thera settlement.

Gwen watched the monitor. "Here it comes. Level nine is the penthouse… eight… seven…" The cab lurched to a stop. "Level six."

They switched on their helmet lights, but when the doors opened, light flooded the elevator from the corridor outside.

"Good job on the lights," Tony said, gesturing for Gwen to step out. The two stood on the elevator platform and studied Gwen's wristpad.

"Pressure's good," she said. They safed their suits, opened their visors. The air was brisk but clean.

Gwen pointed down the hall. "Looks like down and to the left." They turned and followed the corridor for some minutes before reaching a dogleg that led to the central reactor. "Big Bertha, come to me," Gwen said as they entered the chamber. The walls had shielding, despite the fact that fusion reactors gave off little radiation compared to the old fission ones. In the center of the chamber stood a dais. The only light in the circular room came from directly above it, a battery of floodlights that illuminated the platform. The deck was clearly designed to hold something massive, but its flat surface was covered in dust. Cables snaked away from its center, carefully sliced to free the reactor that once held court here.

"Gone," Tony mumbled. "Rhodes is not gonna be happy."

"*I'm* not happy," Gwen said. "Where could it have gone?"

"We know no one from the Martian settlements took it, and there's no record of it being transported to Ganymede or the other Galileans."

"And it makes no sense that somebody would take it the other way. All the settlements from here out were long vacated by the time they shut down Thera and Atlantis West."

Tony scrunched up his nose. "Well, the only thing I'm sure of is it's gone."

"We still have two other chances at nukes, and then we'll head down to the microbial fuel cells on Deck Five."

"If we can get through."

The duo headed down another causeway toward a remote arm of the complex. There, in subdued chrome and flat gray, stood the two small reactors, each the size of a compact car. Faded handwritten signs affixed to their sides read, "Thing One" and "Thing Two."

"Eureka!" Tony crowed. "Are they alive?"

Gwen pointed to the one on the left. "This one's humming. Sounds like happy humming. Let's check just how happy." She keyed her unit and requested diagnostics. "Thing One is operating at fifty percent or so, and it's throttled back, so I think this one's golden." She turned to the other one. In moments, she said, "Thing Two's effectively dead. Problems with some secondary equipment. But the core still looks viable. We could bring back the guts and they could retrofit them back on Earth."

"Sounds like a plan." Tony reached to the side of his backpack and pulled out a tool belt. Gwen did the same.

A motorized hand truck stood nearby, just waiting for somebody who wanted to transport something heavy. They loaded Thing One onto the truck. The lights flickered as the power levels dropped, but power still appeared to be coming from below, from the microbial fuel cells. Getting the core out of the other reactor proved difficult, but by day's end, Gwen and Tony were riding the elevator back toward the surface with their nuclear booty.

"Well, are you happy?" Tony asked.

"Tired, but yes. Happy. Still, I can't help but feel that these are just bandaids for the kind of power Earth needs."

"Maybe there are a zillion bioreactors downstairs."

"We'll know soon."

***

The team loaded Thing One and the core from Thing Two into the cargo pod, a formidable-looking container perched on the ice nearby like an old railroad car. Tony and Gwen overnighted with Taina in the *Belvedere One*. After an early breakfast, they took the long ride back down. Gwen's microbes waited. Hopefully.

The cab passed level six and slowed to a stop at level five. When the door opened, the experience was nothing at all like their level six arrival. They were met with darkness and a nearly submerged hallway. A tsunami poured into the elevator, leveling off just below the control panels.

"This is not so good," Tony said, switching on his suit lamp.

Gwen nodded inside her helmet. "If the bioreactors are flooded, we can't salvage them. But typically, they come in stacks. If the water is halfway up the wall, and the power held in any reasonable way, there might still be a few viable colonies left."

They slogged through the waist-deep waters, turning their suit temperatures up to compensate for the freezing bath. Gwen checked her wrist monitor and opened her visor. Cold air blasted her face, but she wanted to feel the place, to smell it and hear it and taste it. Her breath gusted in clouds before her. The cold of the corridor subdued the smells, but she could sense the aroma of wet metal, concrete, and brine. There was something else, something she couldn't quite fathom yet. She tasted the salt in the air, caught the faint hint of ammonia. This water was fairly fresh from the alien ocean outside. She wondered how long it would be before the level completely flooded.

They reached the entrance to the bioreactor arena, and the smell hit her: rotting detritus. She knew this smell. The microbes were long gone. A wave of vertigo washed over her, but she drew in a deep breath and it passed.

She and Tony searched through arrays of vats, but every one of them had been compromised, most of them years or even decades before.

"I am so sorry, mi amica. Your small friends are gone."

Gwen dragged a glove along the rim of a soggy tank. "It was a long shot, I guess. Maybe at our next stop." But she held out little hope.

She paused, surveying the vats. "Interesting. These reactors are different. They have been modified from anything I've ever seen before."

"Maybe this is evidence of those rumors you spoke of, about new kinds of fuel cells?".

"That's what I'm wondering. Can't really tell from these. Maybe the ones in Atlantis West will be in better shape."

The walk back through the last corridor seemed longer than it had going in. As they passed a light switch, Tony noticed a little sticker pasted over it. It was red and purple. He thought little of it and continued along.

***

"They're on their way to Atlantis West," Chevalier reported to Rhodes. "Got one of the small power plants and the core of a second. Big Bertha's gone, as are the microbial fuel cells."

"Damn," Rhodes grumbled, but he quickly recovered. "All right. Here's hoping for Atlantis West. Any activity out there?"

"Yes."

"Yes? What?" Rhodes roared. He'd had enough bad news for one day.

"Looks like a very small ship coming from Iapetus."

"Another missile?"

Chevalier shook his head. "It's traveling at a leisurely pace, as if it doesn't care who sees it."

"We see it. Obviously, someone's been monitoring Baré's progress. How far is the troop transport?"

"Passed the orbit of Ceres sometime in the wee hours this morning. Picking up speed."

Rhodes rubbed his eyes. "Projection of power. Funny how all throughout history, some things never change. Glad it's so fast. I think we're going to need that military option after all. Send a message to the *Belvedere One*. Advise them of a ship in the area, but don't tell them anything else. Just make sure they keep looking up."

"Right away."

***

"I can't believe we're coming up empty-handed." Gwen's voice crackled in Taina's earpiece. The waves of interference seemed to be getting worse.

"Roger that. No worries. You guys got one good reactor, and they'll be happy to see the core, too." If she was encouraging enough, maybe they could wrap this up so she could return to the sensor panel.

"Thanks, Taina. We'll be out soon." Gwen signed off.

Taina turned her focus back to the radar. Something was approaching. The flight path came directly from Iapetus. Before Gwen and Tony returned, she shut it down and silenced any evidence of the incoming ship.

Chapter 22
You Can't Go Home Again

Taina was still playing the part of the enigmatic woman, oozing with mystique. As soon as Tony and Gwen boarded the ship, she sealed herself in the cockpit with the hatch shut behind her. Gwen could just hear her punching keys and sending some kind of voice message. *Mystique, my ass*, Gwen thought. The captain was getting on her nerves.

Gwen stripped down to her liquid thermal garments, dropping the rigid torso assembly of her environment suit on the floor by her hammock. She wiped a damp washcloth across her face, stretched her neck, and let the light gravity of Enceladus drop her slowly into the webbing of the bed. It felt luxurious to be horizontal after the intensity of the day. She closed her eyes, heard Tony get into the hammock against the opposite wall, and let her mind drift. But the places it went wouldn't bring her sleep.

She sat up and stepped toward the back bulkhead in search of something to read. She opened a small cabinet in the corner. A deck of cards spilled out. She leaned over to pick them up.

"Do you want to play a few palms of something?" Tony asked. "To help us unwind?"

"Hands. Hands of cards. I'm kinda tired. Just looking for something to read."

"Where were those?" He jabbed a finger at the deck.

She gestured with her chin. "Down there."

"I found more in the fore bulkhead."

Gwen was suddenly intrigued. "More cards, you mean?"

"Lots."

"Show me."

Tony lurched from the hammock and opened a small cupboard. He pulled out several decks, each from a different destination: Las Vegas, Syrtis Major, Copernicus Casino. "And then there's this one. It's not really for game, I know. Mysterious?" He held up a Tarot deck.

Taina peered through the front hatch. "The boss likes cards."

"Apparently," Gwen said, glancing at the rear compartment with its separate deck. Suddenly, she was no longer sleepy.

Later, after Tony began to snore and Taina sealed the rear compartment door behind her, Gwen ventured to the comms station in the cockpit. She sent another communiqué to the contact that Melciéna had given her.

Melciéna:
We could use some advice. On our way to Atlantis West. Is there anything we should be watching out for? When was the last time you saw it? How did you leave it? Please respond before tomorrow morning if possible.
Gwen

Gwen assumed that she would not respond in time. Why would she now? With the exception of recent history, she'd been silent for decades. Too little, too late as far as Gwen was concerned. She also knew that if Melciéna didn't get back to her, she would be needing a serious attitude adjustment. Tony would cheer her up. She began a mantra. Forgiveness… forgiveness… forgiveness.

***

Taina took off at first light. The time of day really didn't matter; the *Belvedere One* had plenty of navigation sensors to fly blind in the deepest darkness. But Tony wanted to see the landscape, and Gwen felt that arriving at Atlantis West would be "less creepy" in daylight.

The captain took the ship north along the twisting chasms of Labtayt Sulci, following its curve to the west. The great plain of Sarandib opened before them, with few craters and a surface like a frozen, heaving sea. Low ridges spread in repeating patterns. Somewhere beneath them lay a liquid abyss, deep and vast.

"So bella," Tony pointed, "look at the blue colors, and so few craters. This is some of youngest terrain on entire moon, right here."

The *Belvedere One* drifted west as it descended. Taina followed a band of fractured topography as she made her way northwest across the center of the plain. "Samarkhand Sulci," Taina gestured downward. Bright blue-green slushes emerged from the base of the deepest gorges, flash-frozen in the vacuum. The ice on the plains above seemed warmer, almost tawny in places. A series of compression ridges paralleled each other, rising to form a boundary between Sarandib and older provinces further west.

"That stuff down there reminds me of Europa," Gwen said.

"How brilliant you are, amica. For many of my geologist friends, this is an observation common. The way the ice ripples here again and again, the way the grooves ram into each other in different directions, all very much like some of Europa's neighborhoods. And the scales, they are similar."

As they advanced north, Saturn rose higher into the sky and Gwen became quieter. Soon, she couldn't force herself to give more than one-word responses to Tony's constant travelogue. Eventually, he noticed.

"Something is bothering you, Doctor. I can tell."

"No, no," she said flatly, automatically. She felt his hand on her shoulder.

"You are brooding. It is a good word, *brooding*." He seemed to savor its sound. "It suits you right now."

Gwen closed her eyes, as if to blot out the future. She dug the heels of her palms into them, took in a deep breath, and held Tony's gaze.

He said, "You have been there. There is nothing for you to worry about. Everything that haunts you from your past is gone from there. Empty hallways and quiet rooms, yes?"

"That's all very rational. Even so, I'm dreading seeing some of those places."

"Like your childhood hiding place?"

She nodded. "My sister's secret lair." She shivered. "I can't go back in there."

"It is not where we need to go, is it?"

"It's in the same section, in the same settlement, *on the same moon*. It's just too close for comfort."

"I see it," Taina called. "Hard to port. Almost flew right by."

"Flying by the seat of your pants," Gwen smiled. "I like that."

"That's the way I roll. Two hundred meters. Coming down."

The Atlantis West access was a much larger affair than Thera had been. Several domes formed a central cluster, with passageways fanning out radially to small habitats or outbuildings. Like Thera, the place had a large comms center with several antennae.

Tony put voice to her thoughts. "Taina, why do these sites have such big communications hardware? Overgrown antennas, many backups."

"There's a lot of interference that comes from beneath the crust. Nobody knows why. Sometimes it got so bad that they had to blast their transmissions out. It's been fairly quiet since we got here, lucky for us."

Gwen reached for her helmet. "Let's hope that holds."

***

In a reprise of their visit to Thera, Tony and Gwen hiked across a cryogenically frozen landscape toward the central complex. The ice here was different, polished and slicker. Saturn floated higher in the sky, somehow less foreboding. The Atlantis West access complex was close enough to the equator that Saturn's rings pointed vertically, like a great sword slicing through the heavens.

The airlock control panel was as dead as its sister city's. Tony powered it up, then they entered and cycled the lock. "Here we go," he chirped. "You okay?"

Gwen nodded.

The lock opened to a dimly lit causeway, but the place was far warmer than Thera's chilled hallways. Tony's brows twitched. "Heated? Still? Interessante."

Gwen grinned. "Must be plenty of power. Rhodes said nukes were unlikely here. We must be talking about my little microbial friends. How nice." She flipped her visor open. Tony did the same.

"Yes, good. You have a nice map, I presume?"

"I do indeed. 3D of the entire complex. We need to get down to the seventh level."

"If you were Dante, I would not go with. His seventh level was not a good place."

"No colder than the ice here, though. Shall we?"

They headed for the elevator. Tony punched a key recessed next to the door. Nothing happened.

"Power?" Gwen asked.

He leaned toward the pad. "According to lights, it's fully functional." He stood straight, with an alarmed look on his face. "I think someone has locked it."

Their eyes met. An unspoken understanding passed between them: ignore the implications, get the job done, and get out of here.

Gwen studied the floor plan. "We'll have to access one of the others. Looks like there's a service elevator just to the east. It's the only one near us. This way."

The walkway became murky as fewer and fewer of the overhead lights were glowing. Soon, it was essentially dark, and they made their way using their helmet lamps. In the beams, the walls and floor glistened with a thin coat of frost. The air smelled like a stale ruin in need of airing out, like museums and ancient mummies, decay and devastation. How Gwen wished she could hear those raucous, innocent voices of the Mac III teen team bounding down one of these bleak hallways.

The passages here could have baffled a Minotaur. After a few minutes of walking, Tony became restive. "How long of a trip is this going to be?"

"Thera is pretty small potatoes compared to this place. Thera's spaces are mostly vertical, like hi-rises, while Atlantis West was always meant to be a megalopolis. It sprawls across dozens of acres of seafloor real estate, multiple levels on several interrelated sites."

"I remember the several satellite structures, yes. It's been a while."

"Longer for me," Gwen said, "but I had to study the blueprints for the trip. This service elevator will take us down to the main structure, the one that looks like a butterfly's spread wings."

"Now *that* I remember. And that's where the bioreactors live?"

"That's where all the power is—or was—except for a small reserve power station off to the north that made electricity directly from the ocean currents. Turned out to be such low generation that they resorted to the more conventional stuff as Atlantis West grew. We should be close now."

Up ahead, a glow filtered into the hallway. They soon reached the source: a small ceiling light above the entrance to the service elevator. Gwen activated the button, while Tony studied the panel.

"Oh, looks like it's down there already. It could be some time before it makes it up here. It's on its way."

They both looked down the hallway, back the way they came, their two sets of bootprints black upon the glistening floor. Tony cleared his throat. "So, what do you think could be going on? With the other elevator? Could it really have been in that lockdown mode for almost thirty years?"

Gwen looked down at her boots. "Beats me."

"That's not reassuring at all."

"It shouldn't be. I don't see how the thing could have been in that state for all this time. But everything about this place seems in pretty good shape despite it sitting here, rotting away, for all these years. It's just weird."

Tony thought for a moment. "Maybe Rhodes was right. Maybe there are people here—survivors from Ariel or outcasts from the Vesta conglomerates. We need to be careful. Very careful."

"If somebody's really here, depending on these power plants that we're about to take..." She let the thought die on the frigid air.

Tony thought back to Thera. Suddenly, he snapped his fingers. "And another thing: I spotted a sticker at Thera. Someone had stuck it to a light switch."

"What kind of sticker?" Gwen prodded.

"It was one of those *Embrace your darkening* things. But those came out on Earth..."

"After the impact," Gwen finished. "After Wentaway. Someone's been back here at least that recently."

"Do you suppose Taina has seen anything? Heard anything?"

Gwen keyed her radio. "Taina, come in. Taina, do you read?"

The static fizzed in her headset. "Baré calling *Belvedere One*, do you copy?"

Silence. "That could mean anything."

"The interference seems worse now," Tony said. "Maybe that's it. Or maybe she is taking nap."

"A siesta?" Gwen suggested.

"Spanish term. Italians close shops at midday, but not to sleep. To go home, have some lunch, relax. Then it's back to work until 7 or 8 in evening. We put in as many hours as others. Just split up."

"Sounds smart for stress reduction," Gwen said. "I may take it up."

They waited. The elevator took its time traveling the distance; the oceans of Enceladus were deep and formidable. Finally, the indicator announced its approach, and then its arrival. A tone sounded and light flashed through the opening doors. The two salvagers stepped in. Gwen punched in a command. As the doors slid shut, she thought she glimpsed a footprint just outside the door—a footprint that did not match the pattern on either of their boots.

Chapter 23
Down the Rabbit Hole

Tony

The sea beneath Sarandib was deeper than at the more southerly Thera site, but Tony thought the Atlantis West service elevator seemed more capable that its counterpart at the other settlement. Perhaps the trip would not be so long. The lift's speed rose to a quiet scream as they dropped through the black waters. Unlike the descent at Thera, they had to guess what was going on outside. The functional elevator was strictly business with no windows, and that made things a bit claustrophobic.

Just what Gwen needs, Tony thought. He watched her carefully. Beads of sweat meandered down her forehead, and she wasn't standing still. What was the word? Fidget. Stress scored her face, but then again, Tony felt edgy, too. Atlantis was a far larger facility, and the possibilities of what awaited them down there were many.

The lift did have a full-screen readout, and that readout confirmed that they were traveling at a high velocity. Within minutes, the cab began to slow.

Gwen took in a cleansing breath. Tony knew what must be going through her mind: a return to her past, to the site of her father's death, to the mysteries surrounding Melciéna Valentine and her family history and the fabled Naiads. How would she handle it?

She is a professional, he told himself. *She will be fine. She has to be.* The elevator arrived at a central location on the main structure of Atlantis West, the famous architectural butterfly. The doors unsealed to more darkness, but this murkiness seemed large, as if they stood on a moonless beach at night.

"The place doesn't look like a butterfly to me," Tony said. They both switched on their helmet lamps. The beams disappeared into the darkness. "Not very helpful."

"Must be some kind of staging area," Gwen guessed. "Warehouse, maybe?"

"We should find a light switch somewhere."

"Gotta find a wall first."

The engineers had rooted the elevator shaft out in the open, away from any vertical surfaces, but as he stepped further into the darkness, Tony's light finally fell upon a far wall. He made his way along it until he found a control panel. "Amica, I have it!"

He tapped a command and the chamber flooded with harsh light. Immediately, two of the overhead lamps died in a pop of sparks. The others held. They could smell the acrid tang of burnt insulation and ozone.

"This way," Gwen pointed, not taking her eyes off her handheld readouts. "Oh, wait. There must be another route."

Tony peered over her shoulder. "What's wrong with that one?"

"Don't want to go that direction," she said quietly.

We *don't want to*? Tony wondered, *or* you *don't*?

Tony waited as she searched her database. They were so close now, close to knowing what treasure they might bring back to a power-hungry Earth, close to finishing their mission, close to going home. After a few moments, he couldn't wait anymore. "Maybe we shouldn't be so picky about which way we go in?"

Gwen looked up, seemed to wilt. "Yeah, okay. Here we go."

She led them across the expansive storeroom to an industrial-sized access. Its double-doors accordioned open to reveal yet another dimly lit passageway, this one at an incline. They walked down the slope tentatively, their helmet lights casting surreal shadows. *Surreal*, thought Tony. That's exactly what this was. Salvador Dali would have loved this place. Or perhaps a better comparison was the contrasty, emotionally charged Expressionism: *Metropolis. The Cabinet of Dr. Caligiari*. And the essence of expressionism—insanity, betrayal, terror—what had been left here of those human emotions after everyone had gone?

Any color on the walls was long faded, as if to better accentuate the spooky shadows the two cast as they made their way down, always down. Spots of rust painted ghostly shapes on the surfaces, petroglyphs of spirits long past. Where there had once been carpet, bare floor now met their boots, magnifying the sound of their footsteps. The echoes raised the hair on the back of Tony's neck: creaking walls standing against the pressure of sea water on the other side, the haunting laughter of the compressors and heaters that held the Enceladus deep-freeze at bay.

Gwen paused in mid stride, and turned back to Tony. "We're just about to reach the reactor levels. It's two down, but we'll walk through a short hall on the level just below. Let's pass through quick and get this done, right?"

"Right," Tony agreed. There were things Gwen did not want to see on the next floor down. Could this be where the construction zone was, where she and her sister liked to hide?

Gwen pulled ahead, disappearing around a corner. Tony heard her footsteps on stairs. He followed her down. The stairwell had no lights, so he had to be careful navigating in the narrow beams of his helmet. At the bottom, the stairway opened onto another small staging area with a large hatch. Gwen had left the hatch open. Tony stepped through. He could see her footprints leading to the left. Doorways punched holes in the curving wall, giving access to laboratories and storage. At the far end of the corridor, Gwen stood very still, staring at something down the next hallway. What did she see? Tony rushed to her side.

Gwen's breathing was labored. "I can't go down there. Too much history."

It was obvious to Tony that the poor woman wasn't speaking metaphorically or hypothetically. Her back was against the wall, palms pressed against the cold metal, fingers spread like the legs of a spider. She was terrified.

"You want me to go ahead, check things out?"

"This is ridiculous," Gwen hissed, not moving. "I'm being ridiculous."

Tony couldn't help but feel that she was overreacting a tad. "Gwen, it has been a very long time. Memories are one thing, but should they rule you like this?"

She stared holes into the walkway ahead. "But they can. Memories can haunt a life from across the decades, immobilize people."

"Even after so much time?"

She pursed her lips. "So Tony, I wasn't completely honest with you. I didn't tell you everything."

Tony wanted to be completely clear on what she was talking about. "You mean about your father's death?"

She nodded, closed her eyes, and took a long draught of air. Then she fixed him with her stare. "This is where it all went down. At the end of this hallway. The access is on the right, and it overlooked the construction they were doing with the Naiads. Oh, those creatures were fascinating: graceful and sleek with their long finned legs and webbed hands. And they were fast. Fast with tools. Intelligence of dogs? I don't buy it. Not for a minute."

Tony wanted to grab her by the shoulders, to scream into her face: *We need to focus! We have a job to do. People are counting on us!* But something significant was happening here, and he knew he had to let it play out, for Gwen's sanity.

She shuffled several steps down the hall, keeping her back against the wall. The airlock came into view. It was larger than Tony expected, perhaps eight feet across and deeper than he could see in from here.

"There it is," Gwen said. Her breathing became forced. "I told you how Dad saved me, saved us, got us out of the lock before the window failed."

"Yes, mi amica," Tony encouraged.

"But it wasn't that simple. Dad shoved Claire out the door and gave me a nudge, too. But as soon as we were through…" She paused, perhaps to collect her strength. "I had dropped my diary. My precious diary." She said the last with spite.

Tony remembered Gwen rubbing her thumb back and forth against the cover of the little journal.

Her eyes swam. "And I yelled out for it. *My book, my book!* The water was squirting through that crack, and still he turned around to grab my journal. I don't know what he was thinking. Just one of those things, I suppose. He was on automatic, not weighing the value of a kid's book against human life, ours or his."

"Gwen–"

She had begun to shake. "He should have had plenty of time to get out, but he slipped. He slipped on the wet floor. So, see, Tony? It was me. I killed him. Because of a stupid little book. It's a blame I share with the Naiads. If it weren't for them screwing up the construction site, they wouldn't have called Dad in. The Naiads and me. A team. And then they built what they wanted down here anyway, finished construction as if Dad had never come and gone, and eventually it was for nothing because everybody ended up leaving." She shook her head, rubbed her eyes, let out a long breath. "I suppose it's time to go."

"We will go together." Tony thought about grabbing her hand, just to encourage her along. He reached up for it.

Something crashed at the far end of the corridor. Gwen jerked her hands from her face and locked eyes with Tony. For a moment, she was paralyzed, a mannequin standing in a cramped tunnel. Tony watched as she closed her eyes again. Perhaps she was willing the vertigo and the fear to subside. Maybe she thought about Carter Rhodes, about her sister and the countless other sisters and brothers and family and friends who depended on her, right now, in this moment.

Gwen held up a finger to her lips, signaling for them both to be quiet. She studied her floor plan. Leaning toward him, she whispered, "There's a junction up ahead. I'll go left. You go right and we'll meet up. If anybody's here, we'll see them."

A chill rolled down Tony's back and ruffled the hairs on his arms. "Anybody? *Here?*" he whispered through clenched teeth.

"We just need to be very careful," she said. A flash of distasteful possibilities ran through Tony's mind: refugees from Ariel, settlers from Titan, exiles from Mars. None of them added up to anything good. He hurried down the corridor in the opposite direction from Gwen.

The corridor swung to the left, arcing enough to conceal what was directly ahead. His muscles taut, senses crisp, Tony silently continued down. An access to the right stood open, but it was dark inside. Light poured from another one up ahead. He reached it and gazed in. Empty corridor. He looked down. The frost of the floor had been disturbed, as if something had dragged across it. And on the wall hung the shards of a broken control unit. Whoever it was, he had missed them.

Gwen came up beside him. "Someone's been here," Tony said, pointing to the shards on the floor. "Una persona che è arrabbiata. Someone mad."

"Or someone in a big hurry. When you get nervous, you lapse into your native tongue."

"Hell, yes, I'm nervous. You should be, too. Should we go back up?" Tony asked hopefully.

Gwen crossed her arms. "We've come too far."

"Whoever is here will not be too happy with us taking their power. And what if they are refugees? What if they need help? Some of my favorite people came from Triton, by the way. Was a nice place."

"Our big transport can take them. But I don't think it is. We've seen no evidence of anyone actually living down here long-term. There's no trash, no food wrappers or leftovers. And there's never been any sign of activity on the surface. No, this is something different. This is more likely to be renegade salvagers. Criminals."

"If it is the case of refugees, they will have set up power, which it looks like they have. If it is your salvagers, they'll stop us."

"If they stop us, we negotiate. After all, we have Carter Rhodes and his military behind us. We are here with the World Government's blessing. If this is an illegal salvage operation, we've got him on our side."

Tony spoke through gritted teeth. "So we threaten to take by force? Possession is nine tenths of the law, and we are far from Carter Rhodes. I don't think you have worked this out in your head yet."

Gwen jammed her fists against her hips. "Actually, you're right. This is the first time that I've broken into what everyone told me was an abandoned undersea settlement, discovered the presence of pirates or refugees or whatever else, and been forced to make decisions possibly affecting the fate of the entire Earth. This is all pretty new to me. Got any other ideas?"

"Taina Maes. That is my only idea."

"Yeah, good idea. Backup." Gwen keyed her mic. "Baré to *Belvedere One*. Baré to *Belvedere One*. Do you read?"

A blast of static met her. Tony held up his hand.

"Wait, mi amica. What if someone here is listening? Maybe we should practice radio silence."

"Well, we need to practice something. We can't just stand here, and we can't leave empty-handed. Let's at least figure out what's down there in terms of power and people. It's what we came for. It's our job."

"Right," he said unenthusiastically.

"We're not trespassing. We're investigating."

Farther down, boxes and crates cluttered the passage. One of the boxes had been tipped over, spilling sections of piping across the floor. Tony leaned over and grabbed one.

"What are you going to do with that thing?" Gwen asked.

"Cannot be too careful."

Gwen shook her head at him. "You're a regular ninja." Thinking better of it, she picked up one of the sections for herself. She started forward again.

"That way makes me nervous. We don't want someone sneaking up on us from behind, yes?"

Gwen tapped her chin with a defocused gaze. "What I would give for a bowl full of sunflower seeds right now." She reached down and pulled a small packet from her side pocket. She dropped two pips into her mouth, the taste reminding her of calmer times. Sunflower seeds were another thing that she had shared with her sister. Those commonalities were hard to come by these days. As if emerging from a dream, she said, "We need to be more proactive, take action. If we block the door to the side corridor, no one can circle back on us. We go down, find out what it will take to get any active bioreactors back up, and then bring them topside the other way. Or reevaluate based on the nature of our welcoming party."

"I suppose that is our best wager. Let's do this." Tony sounded motivated, but Gwen had a niggling suspicion that splitting up was not the best idea. She'd seen too many movies. Still—what else could they do?

***

While Tony headed for the hatch, Gwen descended further down the ramp. At its base, the corridor leveled off to a long hallway with multiple access ports. As she passed one of them, she smelled something different, something organic. But it wasn't the smell of moist loam, of the rich soil of a farmer's field. This was something past due, something rancid, something of death.

She peered through the doorway, triggering the lights. Beyond spread an expansive room. Rows of vats lined the long chamber, each with its own light source above. Some held screens and must have been filled with shallow puddles of water for hydroponic farming. Others brooded in the life-sapping darkness, filled with soil. The dirt had greyed over the years, the brown of life long since parched into chalky slate. Something caught her eye. She reached into one of the trays. There, the desiccated remnants of a little flower rested against the side. She recognized it: a strawberry flower. The pink of the blossom had gone to a soft lavender, the green of the stem an ashen gray. She touched it. It crumbled to dust, as the rest of the gardens here had. This was a perfect picture of the cities of Enceladus. Gone were Atlantis West and Thera, turned to dust through simple neglect. How much would it have taken to keep them going? Still, she told herself, the Earth took precedence. The human race. Always the human race and its home world.

***

Tony found some hardware in a storeroom. Amazing what one could do with some plastic, old cabling, and a few short pipe sections. He jammed the conglomeration against the hatch, locking the wheel-like door handle in place. As he spun around to find Gwen, his headset crackled with an incoming message. He had turned it down because of all the background static, and almost missed the voice.

"Dr. Baré, this is *Belvedere One*. Do you read?"

Tony listened to the open channel. No response.

"Dr. Baré, come in. This is the *Belvedere One*, over."

Still nothing. Tony keyed the local channel. "Gwen, are you reading?"

Over his channel, a blizzard of interference came through, along with Gwen's faint voice. "Barely. I'm close to the reactor room and there's all kinds of weird energy fields down here."

"Captain Taina is calling."

After a long pause, Gwen said, "I'm not getting anything from down here. Can you take it? Maybe relay or something?"

"Si, certo," he said. Switching to the other channel, he called, "*Belvedere*, this is Vincenzi. Gwen can't receive directly right now. Can I relay?"

Taina said, "Sure. It's text, so should come through just fine."

Tony waited for the message, keeping the channel open to directly copy to Gwen. Finally, it came.

> Gwen, my dear, Reggie Broadwalter has generously offered to help me get in touch. As you wander the halls of Atlantis West, step gently. You will find out what I mean. Treat all you find as delicate treasure. This is important. It's important to me and important to future events. I will be in touch later, perhaps to help clarify some things. Melciéna

Tony realized he had stopped breathing. He read the message again, knowing that Gwen was doing the same.

How does she know these things?

Tony keyed his mic. "Gwen, what do you make of it?"

Silence.

"Gwen, do you read?"

"Tony, get down here. Now."

He raced down the passageway.

***

Tony followed the convoluted hallways down several levels to the section where the power grid converged. Gwen talked him through the last few turns.

"Make a right after that red bulkhead assembly and you should see the hatch at the end of the hall."

At the terminus of the walkway, a faded label on a sealed doorway read, "POWER ACCESS." The door was unique, different from all the others they had seen. Metal plates fanned out from a central point, forming a flat iris. Gwen stood outside the door, scrutinizing her pad. She didn't look up. "You made it. Excellent. There's power coming out of here, but it's the wrong profile for microbial fuel cells. That's here, too." She shoved the pad toward him and pointed to a wavy line. "Microbes do this." Then she tapped an overlapping line, smoother and higher on the chart. "This is different."

"Like what?" Tony asked.

"Like fusion."

Tony was still processing that news as he said, "But what about Melciéna? The message?"

Gwen twitched. "I don't know what to make of it. I mean, of course we'll be careful. Maybe she's warning us of danger from these people, whoever they are. Maybe she knows something about the structure, like it's unstable or something." Frustration tainted her voice. "I have no friggin' idea. If I put my life on hold every time I tried to figure her out, I'd never get anything done."

"Makes sense, I suppose." But the communiqué from Melciéna Valentine had made Tony as uneasy as Gwen.

Gwen rubbed her hands together. "For now, we concentrate. This is it."

She punched a button on the wall. The hatch irised open, squealing like the sharpening of steak knives. She froze. It stared at her, a giant Cyclops' eye. Tony could see her shiver. A symphony of industrial sounds washed over them as the doorway sliced itself open. Rather than the odor of dust and decay, they were met with the aroma of lubricants and the hum of electricity. There was another fragrance, something akin to yeast. *The bioreactors must be in full swing...somewhere.* It was not what Tony expected, and judging by her expression, Gwen felt the same.

"For once, a pleasant surprise," he said.

They made their way down a row of waist-high transformers. A door at the far end displayed the universal nuclear power sign. Gwen tapped it. "You don't suppose?"

"Any radiation?"

Gwen examined a wall monitor. "Nope. Of course, with fusion the problem is not run-away nuclear piles. The problem is keeping the whole reaction going on its own.

If something went wrong, the reactor would simply be inert. Assuming there is one back there."

She swung the door open. The inner chamber was well lit. Fresh air flowed across their faces as they stepped in. Standing in the center of the room, cables snaking from it like the arms of a starfish, was a large, classically styled reactor with a barely legible sign on the side. Gwen squinted at it and whispered, "This shouldn't be here."

"What does it say?"

"This just should not be here. It doesn't make sense."

Tony leaned over and read the faded hand lettering.

Big Bertha.

He glanced behind himself nervously. "From Thera? But how did it get here?"

"No idea, but it's viable; looks like it's in good shape. Very good shape," Gwen said animatedly. "But we can't get this thing out unless we go through the corridor that we blocked off."

"No problems. I will go around and unjam the hatch."

"I'll get started on powering this thing down and meet you there in a few. I also have to check out the status of those bioreactors." She looked across the room toward a battery of drum-like structures.

Tony jabbed a finger at the fusion power plant. "But do you think someone is using this? Somebody must have gone to a lot of trouble to get it here."

"I'm sure somebody's using it, and if we hang around long enough, I'm sure they'll let us know."

"Do you think this is what Melciéna was talking about?"

She paused for a moment. "I don't think so. I'm not sure what she meant yet. At this point, according to Rhodes, it's a salvage operation. First come, first served. We are here legally."

Her tone brought him up short. "What if there are people here who are counting on Big Bertha to survive?"

"There are millions on Earth who are counting on us to get it back to them."

"Life is not just a numbers game."

She put a hand on his wrist. "Tony, if there are refugees, we'll take them with us, back home to Mars or Earth. First order of business, after we unplug, is to make sure we leave no one behind. A power drop will get the attention of anyone here, and they'll come running. It will be our settlement-wide announcement. The World Council gave us a job to do. Let's go do it."

Chapter 24
Gwen

Twenty of the twenty-four bioreactors were active and viable. They were thriving, in fact. Even after all this time. *How*? Regardless, Gwen didn't have time for mysteries. The fusion system was first priority. Safing the reactor was simpler than she had feared it would be. This one was old, but it had been updated with modern safety protocols, and a few hardware improvements. It would do nicely back home, help to bolster the teetering power grid.

The reactor cables led off into an adjacent room, feeding equipment that was making a lot of noise. As she powered Big Bertha down, effectively announcing their presence, the whine of turbines next door wound down to silence. In the quiet, she heard something just outside the big entryway. She sprang forward and gazed down the darkening corridor. A shadow disappeared at the end. She knew Tony must still be two levels up.

She was petrified, but she knew she couldn't lose this chance to make contact. Josh Aotea's admonition rang in her memory: *Be careful out there, Gwen. It's a big universe.* Grabbing her pathetic pipe weapon, she dove through the passage. She turned the corner and peered down the next segment. The scene was as surreal as a dream, with the same vivid, yet warped clarity never quite seen in reality. The hairs stood up on her neck and arms. A wave of nausea washed over her, nearly buckling her in her tracks. With her heightened senses, everything seemed sharp, impossibly focused. And, as in a dream—or more accurately, a nightmare—Gwen was filled with the sudden intuition that something dreadful was about to happen to her.

There, standing in the center of the corridor, was a creature. Its form was familiar: the reptilian scales, the green, mucous-covered sheen, the huge head, the dark, empty eyes, the slitted nostrils and webbed hands. But this living Naiad—this thing that should have been extinct—was small. Although she had been a child when she first interacted with the subhumans, she knew that they all towered over her mother and father. They were made that way. This one was scarcely a meter tall. It could only mean one thing. But it was impossible.

Naiads were cloned, sterile. Unable to reproduce on their own.

The small being turned on its heel and scampered down the corridor as quickly as it could. Gwen followed. She caught of flash of green disappearing through a hatch. She hurtled through, wondering what she would do if she actually caught up to it. As she passed through the entry, her shoulder clipped the hatch. It clanged shut behind her. As it did, she skidded to a stop. Before her, in this cramped antechamber, sat three large hybrids at a table, with the little one behind them.

Gwen twisted around and grabbed the door release. A low, slurred voice said, "I wouldn't do that."

She turned back, looked around for the voice's owner. There were no humans.

The room tilted. It was too soon in the reactor's cycle for the lights to be dimming, but the place fell slowly into darkness anyway, just as the floor met her helmet.

***

Gwen awoke in a different place. She felt hot, and still dizzy. Perhaps she was suffering from hypoxia. Sometimes, under duress, these suits could malfunction and poison their occupants with too much carbon dioxide, or even too much oxygen. But someone had taken her helmet off. It rested on a shelf across the little room. Next to it lay her little bag of sunflower seeds. Perhaps they had searched her pockets.

She lay on some sort of dais or cot against one wall. The floor was tiled, the overhead light harsh. The room had no windows, so might have been deep in the interior of Atlantis West. Several crates rose in a stack in the corner. Something shuffled behind them. The little Naiad peered out, then leaped to the door. It was gone before Gwen could say anything. So, it was no hallucination.

Where was Tony? She grabbed the headset from her helmet and tried to raise him. Her efforts met only a storm of static.

She stood carefully. The door opened. A hybrid leaned in. She thought it was one of them that she had seen before, but she had a hard time telling them apart. She wondered if this one could speak the simplified sign language taught to many of the Naiads. Not all could vocalize. It opened its mouth, reminding her of an overgrown carp.

"Please, follow me."

So much for sign language.

She couldn't tell where she was, or how much time had elapsed since she'd fainted. Although the reactor's drop in power might have affected the lights by now, the place had all sorts of battery backup, and the microbial colonies were undoubtedly still churning out their juice. She had no concrete idea of her location, no idea of where Tony might be, no backup plan, no options. She could probably outrun this one, but there were more, and she needed to find out what was really going on here. She followed.

Chapter 25
The Downside of Topside

"Relax!" Taina barked. "Take it from the top. Slowly."

Tony was on the verge of hyperventilating. "There are Naiads down there. Live ones. Un sacco. Molti." He waved his hands erratically. "I saw whole lot of them! Moltissimi."

"Where?"

"Through an access window. You must come and help. We have to get–"

Taina grabbed Tony by the shoulders. "Tony, where is Gwen? Did you leave her?"

He pointed absently. "Down there. Down under. By the reactors. We got separated. I panicked. Mea culpa. Have to go get her. I don't know what's going on. Do you have any weapons?"

"Calm down. The first thing we need to do—after you *relax*—is to get a message out."

"To Rhodes?"

She shoved her palm at him. "No, absolutely not. To my boss."

"To Broadwalter."

Taina gave him a derisive look. "Broadwalter, yes. As you wish."

"I think Rhodes has bigger guns."

"Are you going to be okay, Tony? You look pale."

"I am feeling a bit under the climate."

Taina disappeared into the back and returned with a cup of tea. The sugar she stirred into it hid any errant flavors. She handed it to Tony. "Here, drink. It'll calm your nerves."

Tony took the cup in both hands, shaking enough to spill some on the floor. "Graci." He took a tentative sip, then a long draft.

"Good?" she asked.

"Yes." He gulped. "Very. I must calm myself. I'm shaking. I'm dizzy."

Taina watched him, waiting.

"Very dizzy. You know, we must help Gwen. Get back down there." He slurred his words.

She stood and helped him up. "Here, why don't you lie down until you feel better?"

She ushered him to his cot. The drug didn't take long. As soon as his breathing became regular and his mouth drooped open, she made her way forward to the cockpit and flipped on the comms set. The static so commonplace to Enceladus swirled through the speakers. She punched in the frequency code for Iapetus.

***

Gwen's Naiad guide was taller than most. It walked with the same ungainly footfalls of others she had seen, but this one seemed a bit more coordinated than most. She remembered the creatures sliding so gracefully through the water, and then becoming clumsy, seal-like things when they entered into the habs and tunnels of Atlantis West. She also remembered how difficult the verbal ones were to understand. Perhaps that was just a childhood memory. But no, she also recalled Melciéna telling her to be patient when they spoke. Her explanation to the twelve-year-old Gwen had been simple: "Remember how you felt coming home from the dentist? They feel that way all the time." But this one spoke succinctly, with purpose.

By way of explanation, the Naiad said, "I am Thrk."

Gwen waited for it to say something else. Finally, she prodded. "I am Gwen. Pleased to meet you." It said nothing. She tried to encourage a little more conversation. More data. "Where are we going?"

"Below," it said. After a few more dragging steps, it added, "The leader wants to see you."

There was only one leader of the Naiads that Gwen knew of. Could it be?

"Are we going to see Melciéna Valentine?"

The hybrid froze in its tracks. Turning slowly, it rasped, "You know the Vice Gerent?"

Gwen searched her memory. Gerent was some kind of ruler, wasn't it? But more like a manager. And a 'Vice Gerent' seemed even less of a potentate. Sensing an opportunity, Gwen said, "Melciéna Valentine is a relative of mine."

Thrk turned forward again, but continued speaking. "The Vice Gerent is not here."

"Is Melciéna—your Vice Gerent—the leader of your kind?"

The Naiad left a slimy trail on the floor with each step. Gwen had to be careful not to slip on it. The Naiad replied to her question after some hesitation. "She is many things. Counselor. Advisor. Builder. Mentor. But we are going to Kyv. He is the leader here."

The corridor opened into a wide atrium, a place Gwen remembered from childhood. The long, cavernous space made up the core, the body, of the great butterfly edifice, its arching ceiling fanning out overhead like a grand gothic cathedral. The concourse glittered like Enceladus ice, lit from all directions by shimmering beams and shafts of light. The floor mirrored the overhead arches like polished glass. The scene was one not of steel and glass and stone, but of vivacity, of the essence of energy. Every beam, every support, throbbed with the essence of the life force, the fingerprints of forms crafted by living things.

Down the center of the cavernous atrium, a parade of statues stood, mirrored on the floor. Each appeared to be crafted of a different material. Here, a great, porous stone rose like a vertical garden, with vines, veins and flat leaf-like plates. There, another displayed a chrome surface, reflecting its surroundings in tubes and globular protrusions. Next to the soaring support columns, freestanding columns rose from the floor, ending at differing heights. Each was unique in texture or size, and none seemed to have a structural relationship to the edifice itself. These most definitely were not here when Gwen was a child.

"Do these columns serve some kind of purpose?" she asked the Naiad.

"Of course. They are pleasing to the viewer."

Through the floor, the subterranean levels of Atlantis West spread like a twinkling city. Gwen could sense movement everywhere. It was vast. Perhaps what lay beneath her was a substantial human presence, still here after all this time. Why had they kept quiet?

The walls, too, were clear or translucent. From the floor to about Gwen's waist, the material looked like frosted glass, but from the midpoint to the upper reaches of the atrium, the primordial Enceladan sea undulated in floodlights. Outside, the submarine world glimmered with soaring energy towers, vast undersea harbors, complex levels of residential and industrial neighborhoods. Far out in the seascape, she could just make out a flickering light, bright but distant.

"Thrk, what is going on out there?" Gwen pointed toward the light.

"A project. Construction."

Schools of Naiads swam and darted like minnows, swarming in pods as coordinated as flocks of starlings. The underwater ballet was mesmerizing, but Thrk continued on ahead.

Who could be behind all of this? If these creatures were to survive all these decades, surely they could not have done it on their own. Rhodes' idea of refugees from Triton or Ariel was making more and more sense. Somewhere, humans must be tending the great complex, using Naiads to carry out their building plans. But where were they? And what had happened to the creatures responsible for the death of her father? Was Thrk possibly descended from those mindless louts? In outward appearance, they were the same, but these Naiads seemed somehow different—somehow more *aware*—than those of her childhood.

As they reached the end of the concourse, they passed a monumental sculpture, larger than the others. Gwen had never seen anything like it. Its shape echoed the familiar in alien ways. It was at once figural and abstract, with sweeping planes melting into polished plates and fins. It was elegant. And it looked new. Perhaps this was proof of human presence at last.

"Do you consider it beautiful?"

Gwen gave a start. Thrk was looking at her, studying her. She said, "I do, yes. Was it done by..." She stopped herself.

Thrk gazed at the pinnacle of the sculpture. "My grandfather."

There was that Alice-in-Wonderland feeling again, that sensation that every detail making up her past and present might not be the reality she had grown so familiar with. She was nearly back down the rabbit hole again, with the floor vaporizing and the world dropping out, but she fought it back, taking in deep gulps of air. Her sunflower seeds—her security blanket—were still back in her cell.

"Come" Thrk gestured.

What could have happened here? What bizarre twist of biological chance, what perversion of the genetic world, could have visited this place? Melciéna would know, Gwen mused. *If only I could reach her.*

These are not my father's Naiads.

Chapter 26
Cold Awakening

He should have been sleeping the slumber of the innocent, but instead, a blizzard buffeted Tony, a swirling darkness. The room felt as if it was rolling under great waves. His hammock rocked from side to side. He sat up and rubbed his eyes. They felt heavy. So did his head, his feet. He swung his legs over the edge of the bed and carefully placed his feet on the cool floor. With a deep breath, the spinning slowed, the disorientation dissipated.

He reached to the side of the cot and tapped the bedside lamp on. The room steadied. He stood, saw stars. As his vision cleared, he could hear Taina up in the cockpit. Had Gwen made it back? How long had he slept?

Something warned him to use caution, a feeling that Taina's cadence, her tone, was somehow secretive. She spoke in low tones. He couldn't hear the person on the other end of the conversation. He shuffled closer to the cockpit door, which stood slightly ajar. His head pounded, his tongue stuck to the roof of his parched mouth. He tried to concentrate on Taina's voice, her words.

"Oh, he's fine. Sound asleep. I gave him some of that green stuff. Works great… So what about her?...I'm guessing they've probably got her down in the holding area, from what Tony said…I know, boss. But—yeah?... They won't like it. You know how they feel about humans…I share your concerns. And theirs. But what if they want to negotiate? Are you going to go down there, give them some guidance or something? Use strong language? Sounds risky to me…Okay…Okay, right. I'll watch for the ship, boss…thanks. And while you're at it, I wish they could do something about the static. Can you ask? It's getting on my nerves…thanks much. *Belvedere One* out."

Tony felt sluggish, but he was suddenly very awake. He heard Taina shifting in her chair. He spun to return to the back and nearly fell. He lunged into the hammock, reached over to the wall, and steadied it. He tapped the lamp out as the front hatch swung open, flooding the chamber with light. Taina entered, a silhouette looking suddenly like someone he didn't know.

***

Thrk led Gwen down a long passageway lined with bulbous observation ports. Outside, beams of light illuminated the empty blue-green depths. Small autobots crawled over the lines and domes of the exterior trusses and modules. A Naiad swam, mermaid-like, along the outside of the corridor, paralleling their progress along the passageway. She wore some sort of tool harness. For some reason, Gwen wanted to avoid making eye contact. It was disconcerting. Still, she studied the aquatic figure from the corner of her eye. The Naiads of her youth struck her as androgynous, but this one seemed distinctly female, with curves in all the right places and an unmistakable bust line. Another subtle change from the Naiads of old.

Gwen and Thrk passed through an airlock and turned down a short access way. At its end, a massive circular hatch stood like a monolithic barricade, imposing and unwelcoming. She did not remember it from her childhood. She would have.

Thrk delicately tapped a few buttons on a recessed pad in the wall, and the great hatch irised open. Gwen marveled at his actions. The Naiads had always been dexterous, good at fine work and coordination. But they had always needed guidance. Not this one.

As the great lens expanded open, she felt a salubrious breeze wash across her face. She could still taste the salt on the air—it was ubiquitous down here—but there was a freshness to this place. Thrk gestured for her to step through.

Inside, the muffled sounds of machinery blended with the music of a flute. The instrument sounded vaguely South American. It was no recording: the player had a few stops and missteps mingled with the haunting ballad. The Naiads had no lips to curl around such an instrument. The music must have been coming from some unseen human. She wanted to investigate, but Thrk's path took her farther away. She followed the Naiad through another hatch and past the source of the fresh air: dozens of hydroponic trays of vegetables and flowering herbs. The aroma of thyme, parsley, cilantro and basil relaxed her. She took in a deep breath as they walked on. Here were sage and rosemary. She felt like she was in an ancient folk song. But as they crossed the chamber, she caught the unmistakable aroma of garden flowers. Beyond the confined gardens spread a veritable jungle of Mums, Spider plants and Dracaena, plants all rated as top air purifiers. Areca palms, bred for high-yield oxygen release, rose from another section behind. Somebody here knew something about botany. Where were all the human attendants?

Chapter 27
The Hosts of Atlantis West

Tony's head spun, but it wasn't the drugs. It was the series of revelations he had just received from Taina's chat with Broadwalter, revelations that she clearly had planned on him sleeping through. And the way she was looking at him now, he knew that Taina knew. Busted. Both of them.

Tony sat up and rubbed his eyes. When he dropped his hands again, Taina was sitting on the cot across from him, watching. He decided to forgo any pretense of innocence.

"So what was all these things about? That thing about Gwen being in holding area?" Tony could see her jaw tense. She said nothing. "How much you guys know about this place? About what is happening on here?"

Taina looked at the floor, then met his eyes. "When you're under stress, your English goes to hell."

"And how is your Italian?"

She nodded in resignation. "My boss knows the layout of the facilities here. Well. We will work all of this out. You just have to trust us."

"And who, mi conoscenza, am I to trust? I thought you were talking about people down there, when you were speaking. But then you talked of how they feel about humans. Humans? What is going here?"

"There are other parties involved. It doesn't matter. Gwen is in a spot and we can't do anything to get her out just yet."

His head felt fuzzy, massive as a Naiad's. "You spoke of negotiation?"

"Yes. My boss has some influence with those below. He has history with Atlantis West. If anyone can get Gwen out of there, he can."

"You speak like she is kidnapped person."

"No, no. I'm sure we've just had some miscommunication. We'll get it sorted. I promise."

Tony wondered if she was in any position to get things sorted. But before he could press the point, an alarm sounded in the cockpit. Taina jumped to her feet and rushed forward, with Tony close behind.

"What is it?" he demanded.

"Radar. Something big. Looks like a ship has arrived."

"Good. Maybe Carter Rhodes cargo transport has made it?"

"It certainly looks like a ship he sent. The approach vector and velocity imply an Earth source, anyway."

Taina hunched over a round screen, its green glow washing her face with a sickly pallor. Furrows wrinkled her brow. Her comms link pinged. She tapped her ear. After a pause, she said, "Yes, I see it." Another pause, and then, "Great." She tapped off.

"What is it?" Tony asked.

Sitting back from the monitor, she looked at him with an expression akin to panic. "That's no transport. It's a fully armed combat frigate."

"A battleship?"

"And from the looks of things, it may be the vanguard of an entire flotilla."

***

The Naiad called Kyv sat among other hybrids at a low table with several inset screens on its surface. All of them were examining the screens as Thrk led Gwen into the circular room. *Knights of the round table*, Gwen thought. The idea of Naiads studying anything was still a surreal concept to her.

The Naiad across the room, a lithe, female-like figure, noticed their entrance first, but the others followed quickly, tearing themselves away from whatever they were looking at. Perhaps they were playing a VR game, she laughed to herself. That was a thought: Naiads playing a computer game.

Kyv made itself known immediately. Its confident posture and more masculine features set it apart from the others.

"Welcome, Dr. Baré." Its voice had the pinched, guttural quality of the Naiads she remembered, but its speech was quite clear. "Please rest." Kyv gestured to a chair. Thrk stationed itself by the door, crossing its arms over its chest as if it was waiting for something.

Gwen scanned the faces of the Naiads seated around the table, trying to gauge the temperature of the room, but Naiads have no expression to read. She sat.

Kyv spoke first. "We have been watching you since you arrived."

Perhaps the humans had been monitoring the space around Enceladus and spotted them on approach. She wanted clarification. "We? Are there technicians, people down here who help out?"

Kyv rolled its shoulders side to side. It was the closest to head-shaking that a Naiad could perform. "I suppose you speak of humans. There have been no humans here for some time. Aside from the Vice Gerent." It glanced to the young Naiad at its side. "Glf, put your game away."

The gesture added to the surrealism. Before she could catch herself, Gwen laughed out loud.

Kyv scrutinized her. Finally, it said, "The new generation exhibits inappropriate behavior at times." Kyv addressed the group around the table. "May I have some seclusion with Dr. Baré?" The others stood and slogged from the room, leaving Thrk, Kyv and Gwen alone.

Kyv sat across from Gwen, staring at her with those inscrutably dull, black eyes. Gwen did not feel as though she was gazing at a subhuman. This was a being capable of higher thinking...of social nuance...of parenting and computer gaming. Kyv rested its hands on the table, the webbed fingers spread like decks of cards.

"Dr. Baré, why are you here?"

Although everything else had melted into a nightmarish haze, here was something recognizable, something familiar, a solid surface she could put her feet on. The creature was asking her about her role here, and she knew, at least, what that was. But would it—would *he*—understand? She decided to try.

"Kyv, the home world is in trouble. Since the great asteroid impact, our power is waning. Entire power grids are failing, and we lack the resources to get on top of the crisis. We are close, but the majority of our fusion-grade plutonium stores are past their sell-by date—their prime output. Solar power is lower due to the increased dust in the air, and several key power production centers along the coast were wiped out by the blast. Tidal and wind power, nuclear fusion plants, solar, all gone, and all of them supported a very large region. If we can just get some supplemental power, we can get back to a place where we're able to support the various global grids. The solution is twofold: nuclear fusion and bioreactors. The dust will eventually settle and our solar and wind will begin to play the same key roles they did before. But for now, we must have help."

Kyv brushed his hand along his jawline, feathering his gills. "What would you have us do?"

"I was sent to evaluate what energy sources were available here, and take back what I could to supplement Earth's grid. I did not expect the nuclear piles to be so viable, but I hoped the microbial fuel cells were."

"Yes, we have husbanded them with care," Kyv said. Did she sense pride from the creature?

"You have, and that's not always an easy thing to do. It's my area of expertise, so I know how hard you must have worked."

"You saw the ones on the floor with the fusion plant. But we have newer ones three levels down."

"Newer?" Gwen felt a spark of panic. *Who had been helping them*? "Built after the abandonment of Atlantis West?"

Up to this moment, Thrk had remained silent. But he put in, "As you can see, Atlantis West was never abandoned."

Kyv shifted. "You did not expect to see anyone here." He considered her for a moment, seemed to come to a decision. "Would you like to see them? I would be interested in your engineering ideas about our bioreactors. I've been told you tend such reactors on Venus."

How did he know that? Gwen had had it with the cloak and dagger stuff. "Who told you this?"

"We monitor communication throughout the system. Your name appears in many technical writings." He paused. "So I have been told."

Gwen had long considered how she might handle any exiles, refugees, or other humans she and Tony encountered living at these depths. She had planned out strategies for what to say, what to offer them, what leverage she could use to win them over to her side. But in not one scenario did she consider having to bargain with a *Naiad*—let alone one smart enough to outmaneuver her and her team. It was maddening. But she held her tongue, thinking it through. The more she could find out about this place and its workings, the better off she—and the home world—would be. *Knowledge is power*. She tried a different tack. "I would be interested to see your microbial fuel cells. You have been a good host to a traveler like me."

Kyv looked at Thrk, who returned the glance. "You misunderstand," said the creature. "You are our prisoner. Until we can figure out what is going on with the people of Earth."

Prisoner? Things were degrading fast. "I can assure you, their only intention is for me to bring back the power producers that were left here. Bioreactors, nuclear piles. But only what you can spare, of course."

Kyv's tone did not change. Gwen wasn't sure that it could. But something in the room shifted. There was an edge to things now. "What we can spare? We have been warned that this is not the intention of Colonel Rhodes. It seems he wants it all. He cares nothing for whatever might be living here. Nothing for life. You must forgive me, but donating our power supplies to the Earth does not sound as noble to us as it does to you. It would be a death sentence to this place."

"Of course, I understand. But I'm sure we could work out some kind of negotiation, perhaps a trade. You were right when you guessed that we didn't know anyone was here."

"We prefer to keep it that way. Trade is a tricky thing. It is reciprocal. We have a primary rule here about humans."

"What is it?"

"No outsiders. Ever."

A Naiad stepped through the door, looking more like a guard than a guide. It stared at Kyv, waiting. Kyv gestured for the guard to come in. The visitor leaned its head toward Kyv's and spoke in hushed tones.

Kyv stiffened. He glared at Thrk. "So much for your trusting ways." He turned on Gwen. "A ship has arrived. From Earth. Perhaps it is a freighter, but I have never seen a trading ship outfitted with batteries of advanced weapons."

"Weapons?" Gwen was as astounded as the Naiad seemed. "There must be an explanation. Carter Rhodes assured me..." Her voice trailed off.

Kyv turned back to Thrk. "Take her to her room."

Thrk led her in silence. Her hope for human intervention, for negotiating with a supervising intelligence, faded. Perhaps they were telling the truth, and there were no humans here, none of her kind that she could hold a rational conversation with. And what had the aquatic hosts of this place done with Tony?

As they marched into the corridor, Gwen heard the flute music again, just at the next junction. It was lilting, uplifting, the kind of music that soothed the soul and made the heart soar. She knew that even with all their seeming progress in speech and language, these current Naiads had far too little control over their lip muscles to produce such coordinated sounds. Finally, she would see one of their human accomplices, perhaps even talk to them.

As they reached the bend in the hallway, she at last laid eyes on the musician. It was a Naiad, holding the flute to the side of its face, controlling the instrument's timbre with air from its nostril, as coordinated as any lips could be.

Chapter 28
In Trouble and Under Water

Carter Rhodes stood behind the comms technician, surrounded by screens and speakers and various readouts. A massive monitor traced lines of signal strength across its grids. The technician tapped a pad, inputting various commands. He sat back in his chair and looked up at Rhodes.

"It's gonna be tight, Colonel. The interference is at a high level, and we can't filter it out well because of its frequency. It's so close to those we're using."

"Can we go to another band?"

"Not at these distances. And the Ganymede relay is in the wrong part of the sky. Ceres is close. Might be able to use it in a week or two."

Rhodes threw Chevalier a corrosive glare. "What could possibly be the holdup? Something's wrong. They need help."

"Or at least nudging," Chevalier quipped.

Rhodes turned back to the comms officer. "I guess we follow your earlier suggestion, Lieutenant Conners. Boost the signal as far as you can, and we'll try to blast a short message through. Something else, Lieutenant?"

"Sir.., it's just weird. As we shift our transmissions down the spectrum and back up again, it's as if the interference shifts with it. *In response* to it. Sounds crazy, I know."

"Not crazy at all, Connors. The phenomenon is documented. Unexplained, but familiar. We'll just work with it the best we can."

"Yes sir. Ready to transmit burst, sir."

TO: Baré, Gwendolyn
Vincenzi, Antonio
FROM: Rhodes, Carter
RE: Cargo aid
TIMESTAMP: 3:37:04GMT
Have dispatched ship to help with cargo offload on orbit. Understand you may need help. Please advise.

"How's that?" Rhodes asked Connors.

"Perfect, sir. Not confrontational at all. Shall I continue repeat transmissions?"

"Please. Round-trip light time?"

"Just under three hours."

"Let 'em have it for another five or six times, and then maybe again in half an hour. Then we listen and wait."

"Sir."

Rhodes gestured to Chevalier, who followed him out into the hallway. As the door to the comms center closed behind them, Rhodes said, "I think it's just about time to mobilize. Put your team on full alert, Armand. I want them ready to depart for Enceladus at a moment's notice. Depending on what's going on down there, things could get dicey fast."

***

"It seems very simplistic to me," Tony said, jabbing his flattened hands through the air like a judo master. "*Simple* to me. Gwen is somewhere down there. She is circled, surrounded, by Naiads. Not good. She needs help. So do we."

Taina sat there in her high-fashion flight suit, with its severe black-and-white accessories, sounding agonizingly calm. "Carter Rhodes is too far away to do anything but pray. If we call him, he'll just get upset over something that we will resolve here in short order."

"How would we resolve anything? We should ask Rhodes to use that ship of his," Tony pointed at the ceiling.

"Gwen's twenty miles below the ice crust. Their missiles might make a nice swimming pool on the surface, but the Marines aren't going to do much else; not from orbit."

"Right, but we could get them to come down here, yes?"

"Please, Tony. Have patience just a little longer. My boss will show up soon."

"He's nearby?"

Taina smiled. "You might say. Coming from Iapetus."

"All this time Reginald Broadwalter has been cooling his thrusters right next door? Is he with the military outpost?"

"We can either discuss where the boss has been, or we can figure out where he needs to be. At any rate, he can help us more than anybody else can, believe me."

"Then I think you should call him. Right away."

Taina put her hands on her hips. "Actually, the *Belvedere Two* is on its way from Iapetus right now."

"Oh, good," Tony said acerbically. "I'm sure that will solve everybody's problems, and there will be peace in the cosmos."

Taina raised an eyebrow. "You might be surprised."

***

Gwen brooded in her whitewashed quarters, sitting on the cot. She chided herself for the way she'd gotten caught off guard, for her over-eagerness to snatch up those bioreactors and add one more feather to her cap. Her thoughts drifted to her parents, and what they would say in light of her impending failure. What Rhodes would say. And what about Tony? Where was he? Was he okay? Did he manage to escape this whole debacle, or did the Naiads nab him, too?

Some redemptive trip this turned out to be.

The hatch clanged open with a hiss. Thrk leaned in.

"Do you wish some food? You have been here for some time. We can bring you a variety of edibles."

"Actually, I'm famished. That would be nice."

Thrk left, closing the door behind him. Gwen waited for a few seconds, and then tried it. Locked. The room had no windows. It also didn't seem to have any cameras or other monitoring devices. The walls, corners and floors were clean and simple. Still, something disquieted her. Maybe it was the fact that she was thirty kilometers below the surface of an ice moon, surrounded by deadly cold water, imprisoned by green sub-human monsters. Despite the seeming privacy of the space, she couldn't shake that prickling, nagging feeling that she was being watched. *By what? By who?*

She crossed to the far side, where she discovered a small hatch recessed into the wall. In the center of the hatch was a tiny glass button, like the security peep hole on a door. She peeked into it, but could see only darkness. She turned a small handle—it was remarkably cold—and swung the covering open. There, staring directly at her through the glass, were the fierce eyes of a Naiad. The green thing was outside in the water, looking in. It looked like the female she had seen before, the one who paralleled their movements along the corridor. Closer now, Gwen could see its face in every detail. The skin looked soft and supple, not the scaly mien she thought she remembered. The Naiad's eyes, while dull in the air, glistened with life in the Enceladan brine. The female's bearing held a sense of energy. *An intelligence?* Gwen expected it to turn and swim away, but it kept its eyes locked on her.

Here, then, was life. Gwen had met its gaze and found something more than an alien presence. Here, in this strange form, was life born of the same flesh as hers. Disturbing, yes, and evolved beyond her imaginings, but still flesh as her own.

In a few moments, the creature blinked, nodded, and disappeared into the abyssal darkness. Gwen fell back onto the cot, shaking. It was the same sensation that bothered her about Thrk. That's what had made her so uneasy: the fact that he was not behaving like a smart dog or a goon with the intelligence of a three-year-old. She wondered whether something had changed, or whether she'd been lied to all this time.

So the room's hatch was a portal. And others knew she was in here, apparently. She stood again, leaned toward the glass, cupped her hands around the sides of her head and gazed out. As her eyes adjusted, she could make out lights in the distance, long trains of them. Still farther out in the dark brine, the starlike light source continued to sparkle and flash. It took her a moment to realize that the lines of lights she was seeing were actually the far wing of the butterfly-shaped settlement. That soaring gothic spine of Atlantis West, erected on the seafloor, served as the body of the butterfly. Its wings spread open at angles, rising farther and farther up into the waters. Gwen's room was halfway up the northernmost wing, while the lights outside glowed from within the southern one, its surface leaning away from her. The lights seemed, at first, to glow yellow. But as she got used to their luminance, she began to make out colors: crimson and gold, shimmering blues and purples. And within the lights, she saw figures moving to and fro.

This was not an outpost or settlement. It went beyond the bounds of a cityscape. This was a new civilization.

***

Tony waited in the dark. Taina had finally headed to her captain's quarters, and things had gotten quiet back there. Perhaps, finally, she was sleeping.

He made his way forward and closed the cockpit hatch behind him. The fact was that he didn't trust the film noir chick any farther than he could throw her. On Enceladus, you could throw things a long way, but he was fed up and unwilling to play her game any longer. He keyed in the comms code for Earth.

TO: Rhodes, Carter
FROM: Vincenzi, Antonio
RE: Enceladus developments
TIMESTAMP: 14:22:11GMT
There are Naiads here still. They have Gwen. Her situation is unknown. Reginald Broadwalter will arrive soon to communicate with the hybrids. Don't know if he will meet with success. Understand that you have sent one ship. I think we need more.

He took in a deep breath, tapped the send command, and hoped he wasn't too late.

Chapter 29
Civilized Conversation

When Thrk returned, he stepped into the room as if preparing for afternoon high tea. There was nothing threatening in his demeanor, no sense of him being the guard and Gwen his prisoner. He simply set the tray down in front of her and gestured toward it. "I hope you will find some of these to your liking. We were not sure what you normally eat. The various vegetables are fresh, and the porridge has high levels of protein."

He sat down on the room's only chair, directly across from the cot. He stared into her eyes, engaging her as an equal. She had never seen a Naiad do this. It disconcerted her. But why should it? The Naiads had the upper hand here. No doubt Thrk would feel slightly less comfortable in a room filled with humans. If more of her kind were here, he'd lapse into subservient mode right away…

Or would he? Gwen pondered her thought process. She was making assumptions, projections based on past experiences—ones witnessed through the eyes of a twelve-year-old girl. She scolded herself and tried to start again.

The tray held no utensils, nothing to use as a tool. Or a weapon. What it held was remarkable: crisp radishes, pea pods, lettuce, tiny tomatoes, and a cup of something that looked like green oatmeal. She sipped it. It was warm and rich. The spicy taste made its way to the back of her throat and into her nose. It was sharp, but pleasing, bracing in the way hot chocolate could be in the dead of winter. "Thank you for the food. It's lovely."

"If you still want to see the advanced bioreactors, I can take you. They are not far."

"What would Kyv say?"

"It is fine." She thought she could hear a smile of irony in his voice.

"I would like that very much."

"After." He waved a hand at the food.

She ate in silence. How do you make small talk with a subhuman fish-man?

At last, Thrk said, "It makes me happy that you liked the sculpture. It always interests me to see how another will react to something clever like this."

"You must be very proud to be descended from the artist."

"We are all descended from the same creative spirit. I had nothing to do with its genesis, but it pleases me that it has served many generations with inspiration."

Many generations? How many Naiads had come and gone in the interim? And how could she ask?

"Thrk, how old were you when your grandfather died? Do you remember him?"

"Oh, yes. I was mature by then. I was…" he thought for a moment. "Three standard Earth years old. He lived to be eight standard, a very good life."

Gwen did some quick calculations. If the place had been officially abandoned over thirty years ago, and most Naiads lived less than eight years, there could have been four full generations of them since humans left. Four generations. Certainly not enough time for a species to evolve intelligence to this advanced state. What had happened?

"Let us go now. I will be unable to provide guidance later."

Gwen looked down. She had cleared her tray completely. She stood. He opened the hatch and stepped out, then waited for Gwen. As she exited the room, Thrk sealed the access. He walked with her, not in front of her. He behaved like a tour guide rather than a prison warden. Then again, maybe he knew it didn't matter: what chance did she have of escaping a penitentiary buried under thirty klicks of water?

***

"We were warned," Kyv fumed, "by those far more knowledgeable about off-world affairs. Now we see a fully armed battleship. The first human greeting in thirty years. So, Blq, what can we do?"

The female Naiad aide stepped forward. "We've been discussing options. Obviously, while the ship is in orbit, our choices are limited. We can contact the ship. We can cripple its transmissions. We can try to contact Earth. Finally, we can contact the Vice Gerent for wisdom."

Kyv stood and paced. "And if they send a landing sortie?"

"Then our options become less manageable. We can cut power to the elevators. We can also cut off their own communications with their ship or with Iapetus."

"And with Earth as well, correct?"

"Yes. But for the immediate, we want to keep them from entering our seas."

"At the very least," Kyv said. "With the arrival of this battleship, they seem to have increased the risks. It is not wise of them to do. Not at all."

"How would you like us to respond?" Blq fidgeted.

Kyv deliberated for a moment. "For now, nothing. We observe, see what their intentions are. We keep quiet. And await the Vice Gerent's arrival."

"But we are responsible. We are the guardians. We must keep them off the surface and away from the ocean, at all costs."

Kyv leaned toward the aide. "At all costs? You must remember that they are life. Those in orbit above us are part of the creation. They are to be cherished as much as other life forms, yes?"

"Yes, of course."

"Then we will act accordingly."

"Even in spite of the actions they may take against us?"

"We shall see."

***

Gwen stood before row upon row of glistening columns, each a meter across and taller than her. The microbial vats hummed with activity. Her eyes opened wide to take it all in. She realized her mouth was open, her jaw slack. She snapped her lips closed, and then spoke.

"These are not what I expected. They're so tall; how do you monitor the surfaces?"

"There is no need to. The sensors along the side measure for contamination by the electrical levels. They chart the gas levels. The power put out by the microbes is commensurate with their health."

"We've tried that. We can't get the sensitivity we need to tell what's going on at a microbial level."

Thrk simply nodded without responding.

Gwen imagined a scene of these bioreactors installed at Mac IV. No more peering into wide vats. No more opening and closing of ports. Just electronic monitoring. And with their vertical aspect, she could fit twice as many into the same space.

"What's the output?"

"About six times what the earlier versions provided."

"Six times. Remarkable." *Think what we could do back home with this technology.*

"Thank the Vice Gerent. It is her engineering, mostly."

Gwen surveyed the room. Yes, Melciéna, Gwen thought. Her fingerprints were all over this place. It seemed humanity had a lot to thank her for, including the race of creatures now before her. Her engineering improvements were unmatched in the outer system, not least of which were these microbial fuel cells. But at what cost? Who gave her the right to tinker with the Enceladan sub-humans? Didn't she know the dangers of mixing morality into a place it wasn't designed for? When had the autonomous vehicles ended, and the living creatures begun? She was just starting to see the depth of Melciéna's actions.

"Thrk, do your people worship the Vice Gerent as your creator?"

"Worship? Why would we? She is part of the creation. She created us, yes, but she is also created."

The skeptic in her seeped to the surface. "And who created her?"

"Everything has a first cause. Even the universe. When all things came into being, they began as the quantum egg. But an object remains at rest unless acted upon by something outside it, yes? First cause. It is at work even now."

"Very mysterious," she said, and left it at that. For now.

They heard steps coming through the corridor behind them. As they turned, a smallish Naiad plodded around the corner. "Thrk, there is word."

"Word? From where?"

"It concerns our guest. Kyv wants you to bring Gwen to him at the sixth hour." As the Naiad nodded toward Gwen, she realized that something in its behavior had altered: perhaps the line between prisoner and guest was blurring. Just what was her status here, and who was making up the rules?

Chapter 30
Meetings

With the interior lights dimmed and the sun blazing through the cockpit at the front, it was hard to believe that light levels at Enceladus stood at just a percent of those on a summer's day on Earth. This apparent broad daylight was a tribute, Tony thought, to the sensitivity of the human eye. Even so, at the end of the Enceladan day, the eyes felt the strain of living in a dim world. Strange, the tangential things that cross your mind when you are under stress.

"Roger," Tony heard Taina say. She turned in her pilot's chair and called over her shoulder, "Hey Tony, the *Belvedere Two* is on final approach."

"Time to meet the boss," he said, making his way forward. He sat in the seat next to her. The sun hung low in the sky, casting long ebony shadows across the brilliant ice. Within the shadows, Tony could pick out tints of indigo and amethyst, crimson, blues and grays. Here and there, rusty brown stains scattered along fractures, sites of leakage from subsurface waters. Saturn sliced a thin golden scimitar above, the barbs of its crescent pointing away from the sun like the horns of a bull. Its rings shimmered in a razor-straight pencil line aiming directly at the sun. Above and left of Saturn floated an identical crescent, smaller and gray, marking the moon Mimas. To its left, a blaze of yellow light sparked momentarily, and then flashed again, brighter. Landing engines of the *Belvedere Two*.

"*Belvedere One*, we have you in our sights," a baritone voice called over the radio.

"Roger," Taina said. "We're rolling out the red carpet now."

The flashes of light steadied, forming themselves into the sleek body of the descending rocket. Compared to Taina's ship, the *Belvedere Two* was a compact hotrod. Half the length of the ship Tony sat in, it seemed as keyed to fashion as it was to practical space travel. Golden piping outlined the ship's opulent form, with glistening black chevrons on the flanks and small silvery fins for atmospheric entry and exit. A razorback of chrome crowned the spine, wrapping around the cockpit. Any motorhead would be salivating at the sight. Tony was no motorhead, but he had to admit: it was a beautiful ship.

"Thirty meters," the resonant voice on the speaker called out. Tony thought the man belonged at center stage in *Rigoletto*.

The ship descended like a graceful swan, extending skidded landing gear in its final approach. "Ten meters; final flare."

Its engines pulsed brightly for a fraction of a second, and then the chrome vessel settled to the ground in a cloud of ice crystals. Despite the low gravity, the fog around it did not billow or drift, but fell to the ground almost instantly in the vacuum.

It seemed a lifetime to Tony before the airlock opened on the side of the ship. Two figures stepped out, one somewhat thin and the other larger than life. They made their way to the *Belvedere One*. The lock cycled. The hatch opened.

The smaller figure turned out to be a pale, willowy woman with long, salt-and-pepper hair. The larger of the two took off his helmet to reveal a man with rich molasses complexion and hair done in the Caribbean style. He fit his stereotype: bold, brash, and self-assured.

His wide grin put Tony at ease. Tony shoved his hand at the towering man. "Tony Vincenzi. Pleased to meet you, Mr. Broadwalter."

The man took his hand. "Not so much."

Tony looked at Taina, puzzled.

The woman at the big man's side offered her hand. "Broadwalter is a sort of alter ego for me. This is my aide, Jerome."

Taina said, "Dr. Vincenzi, meet Melciéna Valentine."

***

Carter Rhodes let out a long sigh. He squared his shoulders and said, "We are so close, Armand. So close. It would be a crime against humanity to derail the recovery of our entire planet for this small thing. It's a tragedy, no doubt, but we must weigh the outcomes. We must weigh the costs on a human scale. What are these few creatures against a planet brimming with real souls?" He fell silent for a moment, then nodded. "It's time. Send the ships."

"Yes sir," Chevalier said, but his body language said something else. He shifted his weight.

"Well?"

"Permission to speak freely, sir."

"Armand, you don't need to ask. What's on your mind?"

"I'm thinking we need to exercise caution. If Broadwalter can make some inroads with negotiations, force will be unnecessary. We save valuable resources and we look a whole lot more constrained and reasonable."

"All good points. Except you forget that whoever is down there already made the first hostile move."

"Do we even know it was a hostile move? Perhaps Gwen is simply playing the diplomat, parleying with the inhabitants for sharing of their power sources. If we go all cowboy and show up with an armada, that could short-circuit whatever headway she's made."

"True," Rhodes said quietly, weighing his options.

"You and I are military guys. We are used to thinking in terms of force. We're the guys they call when diplomacy fails. But I think diplomacy may be at work right now, and maybe we should see what course it takes."

Rhodes closed his eyes. *What to do?* Finally, he turned to Armand. "No, it's no good. We're weeks out if we send our fastest ships yesterday, and the longer we wait, the less influence we can project. I'm afraid we just don't have the flexibility. It's not me being unreasonable or impatient. It's simple celestial mechanics. Saturn is a long way out. We can't wait."

"I'll give the order."

Rhodes held up his hand. "Belay that. For the moment. It's time I got a bit more involved. I'm going to my quarters to pack. Hold the ships until I get to the staging port."

"You're going out there?"

"Hands-on management. Why not? The *Firecross* is a nice frigate, and I need to get out of the office."

"That's getting about as far out of the office as anybody could."

Within three hours, Rhodes had boarded the *Firecross*, and the fleet had begun its mad dash from Earth toward the ringed giant and its icy moon.

***

Tony gazed upon Melciéna Valentine with fresh eyes. Crow's feet pulled at the corners of her eyes, marking not only the passage of time but also the joys and sorrows of life. Parenthetical creases bracketed her mouth, as if everything she was about to say would begin with *in other words*. Her face held the energy of youth, and her body, though along in years, still seemed to cling to an athletic lifestyle. But it was her eyes themselves that captured him most. Their deep pools of bronze glittered with specks of gold, the whites as pristine as new china. Within those eyes lay the wisdom of the ages. What had they seen? What secrets did they hold?

As Tony processed all this, Melciéna turned her focus on Taina. They sat face to face in the cockpit while Jerome stood in the doorway, inclined against the jamb. Melciéna fanned a deck of cards, absently shuffling, concealing and revealing them at random.

She leaned toward Taina, her elbows on her knees. "We have some catching up to do. Thanks for your good service these last few months."

"Of course," Taina said demurely.

"But now, to business. I am concerned by how things have unfolded below. Our friends are alarmed, and rightly so. Rhodes has a fully armed battle cruiser in orbit." She shuffled with more gusto.

"I'm sure there's more where that one came from."

"Yes. The only thing saving us at this point is distance. It will take the fastest Earth vessels a few weeks to get here. But they've probably got a head start. If I were Rhodes, I would have them en route already, and my sources say that's the way it appears to be playing out."

"Shame," Taina said. "It would have been nice to have some margin."

Tony listened with growing alarm. Finally, he stepped forward and stuck his head into the cockpit. "It sounds to me like you guys aren't too worried about Gwen. So whose side are you on, really?"

Melciéna frowned, tapping the edge of the deck on the console. "It is a complex issue, more complex than you think. We are not necessarily against anyone. And make no mistake: I love Gwen. She's my goddaughter, after all. But there are things here, on this little ice moon, that must be sheltered, protected from outside influence."

Tony's eyes burned holes into the duo. "You'd sacrifice her and everything else to help some human rebels keep their little fort?"

Melciéna narrowed her eyes. "The only human down there is Gwen. The power now lies within the hands of the Naiads."

"The Naiads?" Tony thundered. "They're simple tools; somebody else pulls the strings, yes?"

Taina said, "You may have jumped to some unfair conclusions based on false information."

"Very possible," Melciéna added. "Of course, you wouldn't be the first."

"Well, as the mother of these genetic throwbacks, maybe you can enlighten me." Tony tried to curb the anger in his voice, to no avail.

Melciéna stood, clasped her hands behind her back, and gazed out the windows. "Perhaps you deserve a little history? Why don't you sit, Tony."

Her voice was gentle, controlled, but it still felt like a command. He entered the cockpit and settled grumpily into the third chair. She kept looking outside, but Tony got the impression she was looking farther than anything in that frozen wilderness. She was looking back in time. He studied the woman, starting to understand why Melciéna Valentine's reputation had grown to mythic proportions. Even just standing there, contemplating, she was a force of nature. She moved with certainty, with an intensity constrained only by grace, sinewy and athletic and severe. The word that came to mind was formidábile, *formidable*. He was more comfortable with her looking out the window than at him.

"Most people know, on some level, that the Naiads were the product of illegal genetic experimentation. This is not a secret, but it's something that's often played down in the historical narrative. Those who created them in the early days of Atlantis West could probably have gotten away with it had the place not expanded so quickly. But Enceladus became a cosmopolitan center for the outer system, and their activities were uncovered. I was brought in to clean up the mess left by the criminal geneticists, who were, by then, in jail."

"As they should have been," Tony added.

"Granted. But the Naiads they had created became a problem. What to do with them? The Enceladans here decided to take them under their wing, train them to build and repair, not as slaves but as servants. They were to become productive, working members of the colony. That was the grand plan: help the Naiads to become smart enough that they would never again be relegated to a slave force."

"I'm surprised, of course. Maybe happy for them. They certainly seem to have progressed to a place where they have their own—what do you say?—agenda. But your new race of Michelangelos are holding Gwen hostage."

"I wouldn't worry about her," Melciéna said, almost dismissively. "If I know Gwen, she's probably staying up late down there, burning the midnight biomass. Exploring. Tinkering." She held a small metal hypowrench up in front of Tony, smiled, and made it vanish with a flourish. A moment later, the wrench clattered to the floor behind her. "Oops. I'm a little rusty."

"I think you should be more worried about Gwen than you are. They've kidnapped my colleague. What could they possibly want?"

"The latter is the important question. We should worry ourselves not so much on the 'kidnapping,' but rather the motivation." Melciéna turned to Taina. "Whatever the reason, their motives are not as questionable as those of the large frigate sharing our orbit. I think we need to have a chat with that ship. Do you have the frequency, Taina?"

"Yes, stored on my console."

"Raise them now. Ship to ship direct."

"Yes, Melciéna."

The comms speaker flooded the cockpit with the Enceladan static, but the interference faded as the other ship came on.

"*Belvedere One* to Earth cruiser. How read?"

After a pause, the reply came in. "This is the *UFM Chronos*. Read you five by five. Please clarify your identity."

"This is Reginald Broadwalter's personal yacht, commissioned by Colonel Rhodes to aid in operations out here."

A longer pause ensued. The voice on the other end said, "Roger that. I see. Mm. How can we be of assistance?"

Melciéna pointed her deck of cards at Jerome and swept them toward the microphone. He leaned into the comms set, adopting a formal tone. "This is Reginald Broadwalter. We are in the process of some rather delicate negotiations, and request that you stand down from any planned landings at this time. This is, of course, a request, not a demand. But our request is in keeping with Colonel Rhodes' mission for us."

The customary pause stretched on. The reply came in a businesslike tone: "We will contact headquarters and advise you of your status. *Chronos* signing off."

"Well," Taina said. "That's that. They aren't very conversational." She reached over and punched Jerome in the shoulder. "Hey, you make a pretty good Reginald."

"Happy to do the voiceover, but let's leave the real persona to Melciéna."

Melciéna stood. "There's no way around it. We need to go down. I'll take Tony. I need you two up here on watch. Carter Rhodes' cruiser is making me nervous."

"Fine with me," Taina said. "I hate the cold."

Melciéna smiled. "Not as cold as last time you were here. For that matter, it's not as cold down there as it is up here."

"Not in my cozy little cockpit."

Tony wondered just how much history this shipload of people had with the green ghouls below. He guessed he was about to find out.

Chapter 31
A New Light

Thrk unsealed the hatch to Gwen's room. They entered and sat. In the close quarters, Gwen could smell the scent of the Naiad, a salty fragrance, not fishy in any way, but definitely aquatic.

The Naiad was silent for a moment. He finally put words to what Gwen was thinking.

"Something has happened. Something at the surface. Or in orbit. This is not good."

Gwen's mind raced. If someone else had arrived on the surface, they would be coming directly for her, or for the energy sources. It could only be good news for her. But she was beginning to doubt. The nice, neat roles she had envisioned for herself and her team were blurring like the lines on a watercolor as more players and more layers were added to the mix. Where did they stand? Where did she?

Thrk glanced at the peephole in the wall, then continued, "We have dedicated ourselves to the preservation of this world, to the protection of its waters. Anyone on the surface will not know that. They will not know what is at stake. This has been by design."

"I don't understand."

"That is good, Gwen. That is as it should be."

"Why can't we just collaborate on this? You guys give the humans some of your power, and we agree to leave you alone? No bad guys; no good guys. No light versus darkness."

"And this Carter Rhodes would follow your advice? Keep the humans off the surface?"

"I think I could convince him."

"We are not sure he is to be trusted. He is a man against the wall."

"He's not such a villain."

"Perhaps darkness and light coexist in us all," the creature said.

And just like that, staring at the creature beside her and mulling over the words it had just spoken, the very essence of Gwen's world changed. These past few days, she had been treading lightly along, trying to make sense of this twisted path. It was as if she suddenly realized that what she had been walking on all her life was the ceiling, and she had to fall in order to find the new floor on which to make her way. Her ears rang. Her mouth parched. She needed something crunchy.

Thrk's rheumy, round eyes gazed at her from their deep pools of gelatin. "Yes, you see it, don't you? Life. Spirit. In you. In me. If the breath of the Creator is in us, are we not holy? Are we not the same flawed things, finding our way in a universe far too mysterious for any of us to truly comprehend?"

Gwen smiled and scratched her head. There was more Socrates than servant to Thrk. "I suppose I've been bumbling through the universe most of my life."

"Yes. Every sentient being has, I think. Should we not, then, break bread together at the same table?"

She wanted to say something, to agree on some level, or to at least explore Thrk's remarkable thought. But the words wouldn't come. She had crossed the void to this strange ice world, to meet another sort of humanity. In the Naiads, she looked upon flesh unlike hers, but flesh that flowed from the same fount. Here was life. Here was spirit. And woe to anyone who did not find it holy.

Thrk stood. "It is time. Let us meet with Kyv."

As they traversed the great atrium, Gwen saw everything in a new light. The statues, the columns, the revamped floors and walls: now she could see how it all was perfectly engineered to echo the aquatic environment outside, rather than serving as a bastion against it. The undulating flow of line, the triumphant rise of the cavernous ceiling above a multitude of reflective surfaces—every feature contributed to something alien and, at the same time, familiar, a place with the essence of the aquatic.

As they entered the Naiad nerve center, Kyv remained seated, tracking them with his eyes. Behind him, a map screen showed three blips, two next to each other on the surface of Enceladus, and one in orbit. Thrk and Gwen stood before Kyv in silence.

The Naiad leader sat forward, his air-dulled eyes active, intense. "Someone has arrived. Someone from off world."

"We suspected as much," Thrk said.

Kyv lurched forward further, his face nearly meeting Thrk's. "Did you suspect that we would have a fully armed battle cruiser over our heads? Did you suspect that the armies of Earth would be coming our way?"

"I had hoped not."

Kyv sat back, relaxing his shoulders. "I had hoped not as well. But here we are."

The tension in the room was palpable.

Kyv turned to Gwen. "How did you like our bioreactors?"

To those around them, it might seem that Kyv was changing the subject at an inopportune time. But he was not. He was focused on the crisis at hand, and Gwen knew it.

"They are beautiful," Gwen said. "Advanced beyond anything we have on Earth or Venus."

"Or Mars," Kyv added. "Yes. And do you think this technology would interest Carter Rhodes?"

"No doubt. I would love to show him."

"And do you think you might convince his people to remain in orbit, rather than coming here, if we could ship one up for him to examine?"

"I consider that likely. Look, Kyv, Rhodes wants only one thing: to add enough energy to the Earth's power grid for it to reach critical mass again. What if the Naiads of Enceladus agreed to manufacture these biological power plants for Earth? In return, humans could agree to land in only pre-arranged sites for pickup."

An aide stepped over to Kyv and whispered. Gwen could just make it out: ...*in the elevator...descending.*

Kyv turned his attention back to Gwen. "But what would we, the inhabitants of Atlantis West, gain in return?" The Naiad was wily.

"I guess that depends on what you want. Or need. With enough energy and time, we can work anything out. For both of us."

"Good words," Kyv said.

Gwen replied, "And you and I will make those words live."

"I will seek council with those who have arrived."

She and Thrk looked at each other. Gwen said, "Arrived?"

The main door squealed open like the maw of a great carp. A hush fell across the chamber instantly. All eyes—human and Naiad—turned to the door as it flared wide. Light flooded in, and within the glare stood two silhouetted figures. As the door constricted to a close behind them, in stepped Melciéna and Tony.

"I knew it," Gwen said, remaining at Thrk's side. A little distance between the outsiders and Naiads seemed to be what the Naiads wanted. At this point, they were possessive of Gwen, kidnapped or not. "I knew it was you from the moment we set foot on that ship."

Melciéna raised a brow. "Was it..."

Gwen nodded and pursed her lips. "The cards. Did you leave them for me?"

"It was the only thing I could think of. Nice detective work, Gwendolyn."

"You knew?" Tony exploded. "Gwen, I have been worried ill. Why you didn't tell us you knew Melciéna was part of this?"

Gwen moved toward Tony, but when Thrk tensed, she stayed on the Naiad side of the room. "I'm sorry, Tony. So sorry. I just wasn't sure. Not until you all walked through the door."

"Still would have been nice to share."

"Regardless, I appreciate you going for help," Gwen said.

Tony looked down at his shoes. "Actually, at the time I was thinking of me more than of you. I panicked. I ran. I'm sorry...Will you forgive me? This is my first time dealing with Naiads and power plants and military exercises all together."

Gwen grinned at him, thinking back to their earlier exchange, before they knew what lay in store for them. "Of course I forgive you. It's my first time, too."

"It seems we have much to discuss," Kyv said, quieting the room. "You may 'catch up' later. We will host a welcoming meal." Kyv nodded to one of his attendants, who left immediately. Then he stood and walked as briskly as a Naiad could toward Melciéna. He reached out his hands to hers as they met in the center of the room. "Welcome, Vice Gerent."

She took his hands warmly. "Are you well?"

"Yes, yes. And you are safe and healthy?"

"As a horse. Do you know that idiom, Kyv?"

"I did not. It is as obscure as most of the others."

"Suffice to say that I am healthy. After dinner, may I visit our bioreactor experiments?"

"I will take you myself. Our microbes are also healthy as a horse."

"Excellent." Melciéna turned to look across the room. "Gwendolyn." She held out a hand.

Gwen crossed the chamber, stopping several feet away.

Melciéna dropped her hand. "I've been following your activities here from a distance, as best I could. What progress have you made? What agreements have you come to?"

Gwen held up her palm. "First things first: why did you keep all of this hidden? And what part did Reginald Broadwalter play in all this? The rescue from the cycler, the transport to Enceladus–"

"There is no Broadwalter," Tony said.

"Is too," Melciéna objected.

Tony pointed at Melciéna. "She is Broadwalter."

Melciéna dropped her chin slightly, admitting the truth of the matter. "I had to step in. When it became obvious that you were coming to the outer system anyway, even after my warnings, bringing who-knew-who, I decided that I needed to keep a closer eye on the situation."

"You mean control the situation."

"As you wish," Melciéna conceded.

"So you sabotaged the cycler?"

"I would never endanger the lives of passengers like that. No, the ship did that on its own. These cyclers are ancient, you know. But when I got word that your ship was crippled, I saw an opportunity. Taina—my accomplished pilot—was aboard already, keeping tabs on you and Tony. With her talents and training, it was easy to have her standing by for the *Belvedere One*. Worked out well, don't you think?"

"You haven't changed, have you?" Gwen hissed in an even, menacing tone. "How dare you."

Tony flinched.

Melciéna's smile froze and faded. "You mean, how dare I arrange for the rescue of twenty seven passengers and crew on a crippled cycler? How dare I arrange for you to get to Enceladus in an almost impossible situation? How dare I protect you from Carter Ray-gun's big ship while you scoped out two sites on this little ice ball?"

Gwen's fists were on her hips. "Listen to all that arranging you did. If our roles were flipped, wouldn't you feel the least bit manipulated? Used? Taken advantage of?"

"Oh, I don't think so. I think I would see it for what it was: a pragmatic approach to a series of problems."

31 A New Light

"And while you were being pragmatic, did it ever occur to you to share your bioreactor technology with the suffering masses of your own species on Earth? You nicked out of the game right after they abandoned this place. Why?"

Melciéna's cool smile returned. She looked around the room. "Aren't you getting some idea of what changed here? Of what I needed to do? You're looking at a new race, beings that needed guidance to survive. You and I had the same goals, yours with the human infrastructure, and mine with the Naiads. Survival of a species. Can't you see that?"

Gwen could see it. Yes, she could see it all now, but she was angry. In the moment, she still felt betrayed. She said nothing.

Tony broke the silence. "There has been a lot of water over the bridge since we talked last."

"There has indeed," Kyv said, taking control of the conversation. "Colonel Rhodes has sent us a gift in the form of a large cruiser with large weapons."

"One thing is clear," Melciéna said. "He cannot be allowed to deploy a landing party."

"Agreed, Vice Gerent," Kyv said. "We have discussed several options. But we are unsure of what course to take."

Melciéna addressed Kyv. "Are you under the impression that the ship's crew are taking orders from their sources at Earth?"

"We are under that impression, yes."

"As am I," she said, a glint in her eye. "And what might we do to buy ourselves some time and make it necessary for those in that vessel to speak to us rather than to Earth?" She spoke to him as a teacher would to a pupil.

Kyv said, "One of our options is to cut off their contact with home."

"Yes, good. I thought so, too. But the timing is important. It must be done before they attempt a landing. We do not want to cut off communication between the orbital crew and the landing party. That might endanger them."

"I see," Kyv said. "Shall we make it so?"

"It might be prudent. We will contact the ship at an appointed time, a window in which we choose to clear the interference."

Kyv stepped over to Gwen, stared into her eyes for a moment, and then turned his gaze upon Thrk. "Thrk, please make arrangements."

"Certainly, Kyv."

Kyv met Gwen's gaze again. "Would you like to take Gwen to the alcove with you?"

"I will need to concentrate. It will be delicate."

"Understood. Gwen, you must remain here. Perhaps you and the Vice Gerent can do your 'catching up' now."

Alcove? Gwen wondered what that could mean, but there was no time to ask. It seemed that Kyv had dismissed Gwen and the others. The Naiads standing guard now shuffled aside. Melciéna told Gwen and Tony to follow, explaining simply, "I'm in the mood for a stroll."

As the trio sauntered down a central corridor, Gwen pondered the events of the last few minutes. She had seen no communications equipment in the Atlantis West complex, although it was a big place. Her curiosity got the best of her. "How would they do that?"

Melciéna's pace didn't slow. "Do what?"

"How will the Naiads jam the transmissions from Carter's ship?"

"My dear, Enceladus has much to teach us all. Many secrets."

As Melciéna turned a corner, the path took on a sense of familiarity for Gwen. Goosebumps scattered over her arms; a chill rolled from the base of her skull down to the tip of her spine, then down until it settled like a stone in her gut. "Where are we going?"

"I have a favorite spot. To think."

"Why don't you guys slow down a little?" Tony called from behind.

Gwen needed no prompting. She stopped in her tracks, rigid. "Don't."

Melciéna dropped her voice nearly to a whisper. "Gwendolyn, you need to go down here."

"Not there. There's nothing for me there."

"Nothing but bad memories? But if you try, you may find something more. Please, do it for me. And if not for me: for you."

Melciéna put her hand on Gwen's elbow, gently guiding her. Together, they walked a few more steps, turning into a small observation deck with an airlock at the end.

"What is this place?" Tony asked.

Gwen took in a slow breath. "This is where my father died."

Chapter 32
Interference

"Ineffectual." Commander Chang-Lopez of the *UFM Chronos* scratched the back of his neck.

"Completely." The comms officer was adamant. "We've boosted signal as high as it goes. On a scale of one to ten it's about a zillion. Nothing."

"And you tried contacting Iapetus? Right next door?"

"No response, sir. Nothing's getting through."

"How long have we been out of contact with Earth?"

"Seven hours and change."

"That will trigger some action on their part soon."

"Yes sir. Another hour of silence from us and they'll start actively searching."

"Sure. From a billion kilometers away. Not much they can do in the short term."

The comms officer said, "One of their first action items will be to contact Iapetus base and have them make some attempts. Iapetus is close enough, and they've got enough power, that they just might get through this mess."

Chang-Lopez chewed the inside of his cheek in thought. "Can we deploy one of the comsats as a relay?"

"Already tried, sir. And failed."

"All that background radio static." He shook his head. "It's not even this bad at Jupiter. It's the damnedest thing. And you're sure it's not coming from Saturn's magnetosphere?"

"No, sir. I mean, yes, we are sure. No, it is not. It's coming from inside Enceladus. And it seems rather specific to our common frequencies. As if–"

"As if there's someone down there who doesn't want us talking to Earth?"

"Yessir. I guess so. Sir."

"Well, then, that leaves us only one option. We need to send a recon party down to find out what's going on. An armed party."

"Sounds reasonable," the comms officer encouraged.

"All right. Then that's what we'll do. Keep monitoring the radio spectrum. And try to set up a way that we can talk to our own people who head down into harm's way."

"Yessir." The comms officer left, wondering just how he might carry out the Commander's orders.

***

"Gwendolyn, losing your father hurt us all. He was a good man. Clever. Funny."

Gwen's eyes burned. She fought back the tears, steadied her voice. "Yes, I know. I grew up with him."

Melciéna said, "Don't forget, my dear, I knew your father longer than you did. That twitch he would get in his eye whenever he was nervous about something. The way he tipped his head when he was trying to understand something."

"Like a puppy," Gwen agreed. "Yeah, but when it comes to total hours, I got you beat."

"Granted," Melciéna smiled.

Tony lingered in the doorway, not knowing whether he should enter or disappear down the corridor.

Melciéna looked out the window, at the starlike light in the distance. The undulating, gauzy radiance of the Enceladan sea washed across the ceiling of the airlock, scattering golden polygons down the walls and onto the two benches facing each other. Looking back over her shoulder, Melciéna said, "What hurt even more was watching you punish yourself for it, year upon year. My dear, you have paid any price owed, many times over. But you did not cause anything. We all make decisions. Your father decided to turn back when he should have followed you out. He died, not because of your childlike request, but because of his own decision. He was the grownup in that moment, not you."

Gwen wanted to believe it. She wanted to be absolved of the guilt, the weight of the years. But all that regret was building in her gut, rising in her chest, pushing out the air. "I wish it was that easy," she said, embracing all the pain and the remorse that she had collected since she was a child of twelve. It was a formidable collection, one that continued to grow. After all that stockpiling, how could she give it up?

Melciéna turned to face her. "Gwen, look at me. Of course it's not easy. There are times when the very worst happens in life, and you can't let go of feeling responsible, even when you know—logically—that whatever happened was not your fault. But in the end, it's not about logic. There is always that emotional component that dogs you, makes it tough to just let it go. But shame is like glue. It sticks to us and binds us. You don't need that, do you? You need to be free of it. And given time, you can let it go."

"It's not shame. Not really. Sometimes the angst of a situation can inform your actions, show you the way. Inspire you to do the better thing." Gwen could tell Melciéna wasn't buying it. She didn't know if she did, either.

"Gwen, you are still trying to measure up. Still trying to be the daughter who could save her father. But you couldn't save him, so now you're trying to save humanity. You got that from your mother, too. And while you're at it, you're still trying to be the big sister that Claire needed in the moment, even though in that moment, no super-sister would have done. And although you are estranged—I assume you still are—you carry the guilt of not being there for her, the ever-dependable big sister. And that deplorable diary."

"What's that got to do with anything?" Gwen asked with more force than she intended.

"I really think that little diary of yours is what led to your ultimate sisterly split. The more you put your deep twelve-year-old secrets in it, the more I saw Claire come to resent it. It was innocent, I know; you were young. I suppose she felt threatened, as if it was taking her place on some level. It's something you normally would have grown out of, if it hadn't played such a role in your father's death. But a small thing, an odd event or a seemingly inconsequential object, can turn a little crack into a wide canyon."

"Hard to be big sister when there's a canyon to jump." Gwen rubbed her temples.

"Equally hard to save the human race from itself. You are strong and smart and accomplished. Why don't you forget about measuring up to some arbitrary standard? Goals like that can be moving targets."

"Life keeps reminding me." Gwen felt the weight of the diary in her pocket. It was a weight that surpassed the gentle gravity of Enceladus, the weight of passing years and lost dreams, a heaviness shared with her sister in a way no one else could share it.

"I remember my mother telling me I would always be the big sister, no matter how old I was." She pulled the tattered journal from her pocket and studied the cover. "Frankly, it isn't fair of me to hold on to this. Not fair to those around me who love me."

"Nor to those who you love," Melciéna said gently.

Her godmother was right, she had to admit. But the void her sister had left in Gwen's life was deep and ragged. The break in their relationship was like the death of her father. Like the death of a friend. Claire had always been there for her—braiding her hair, scaring the hiccups out of her, painting her nails bright purple, sharing her sunflower seeds. It hurt to think what might have been. The rift widened as time soured the experiences in the airlock, made the secret of Gwen's journal rancid, a weight nearly too heavy to hold. No one knew the details of Hugh Baré's last moments except the two sisters and their confidant, Melciéna. And now, Gwen could add Tony to the list.

She placed the diary on the windowsill but kept her palm over the cover. "Of course, there were other factors involved," she said slowly.

Melciéna shifted, as if the conversation had suddenly reached a new level. "So we come to it, don't we? *If the Naiads had not failed in their construction techniques, there would have been no structural collapse, no crisis for your father to attend to.* Yes?"

Gwen felt the oppression on her shoulders. She slumped under its weight. She remembered the screaming glass as the water jetted through, and the Naiads looking in helplessly. Another image—a vision, and not a memory—came to her then: Atlantis West without its microbial fuel cells, without its nuclear reactors, without anything alive. The flowers, the vegetables, every Naiad, all turned to gray dust like the flower she had found. The image of the Naiad child flashed across her mind. "Ironic. I've been tasked with something that will lead to their destruction, and I can't seem to do it."

"Perhaps they are not the monsters you thought? Perhaps you are finally assenting to the fact that the Naiads who stand before you are no mindless things responsible for the death of Hugh Baré."

Perhaps, Gwen thought. Perhaps, all this time, she had harbored anger, even hatred, toward beings who didn't deserve her wrath. But that wad of self-shame had made it to her throat and she had no voice.

"Maybe I should go," Tony said, turning in the doorway.

"Maybe you should stay," Melciéna said, flicking her long hand toward him.

Melciéna stood and squeezed Gwen's shoulder. "These are big questions, large issues that span a lifetime. You will need time to figure it all out. But take the time, Gwendolyn. Even in the midst of life swirling around you, take the time. Ponder a little. You owe it to yourself. And to your sister."

Gwen's sorrow turned to anger. After all, Melciéna had her own actions to answer for over the years. Where had she been when the people of Earth needed her? Huddling in the outer system with her pet sea creatures? As Melciéna turned to leave, Gwen found her voice, found the will to ask the big unknown of her godmother. "Why did we make them?" As soon as she put the question to voice, her anger melted, turned to something softer. Gentleness had crept into Gwen's timbre, a desire not to confront, but to understand.

"Them? The Naiads? Such questions." Melciéna pursed her lips. "Efficiency. Economy. When you set up AI in a glorified submarine, it's–"

Gwen put her hand on Melciéna's arm. "Not the party line, Melciéna. We know all the rationale." She lowered her voice, met Melciéna's eyes, and asked again, "Why did we create them?"

"Because we could, I suppose. It's artificial evolution, a natural progression of things."

"More like regression, wouldn't you say, Dr. Frankenstein? Sub-humans with IQs below sixty–"

Melciéna stopped her, bristling. "They are *not* subhuman. Certainly not. Different, yes. Not less. They're not as imbecilic as Carter Rhodes would have you think."

Tony squirmed. "But how could they have survived on their own? All this time?"

"I–" Melciéna looked down at her interlaced fingers, struggling. "They broke free."

"But they weren't designed to think that way," Gwen objected. "How did it happen? What went wrong?"

"Wrong?"

"Or right, as the case may be?"

"I helped them."

Gwen straightened. "You. Helped. Them. What does that even mean?"

"You see, it comes down to a numbers game. Numbers of neurons," Melciéna said. Meeting only judging silence, she continued, "Neurons are the information processing units of the brain. Most of them are clustered around a nice little neighborhood called the cerebral cortex. That's the part of the brain that handles logic, abstract reasoning, and so forth. Even humans like Carter Rhodes have more neurons in their cerebral cortex, compared to the number of neurons in their entire brain, than any other creature. In short, it is the cerebral cortex that separates us from the creatures roaming the African savannah. Elephants have huge brains, three times the size of ours, but a relatively small cerebral cortex."

"I see." Gwen said it hesitantly. Tony looked completely lost beneath the onslaught of these new terms.

"So to the numbers game. We humans have 86 billion neurons, and 16 billion are dedicated to that all-important cerebral cortex. Lowland gorillas have half that, despite the fact that their brains are larger than ours. As for the Naiads, they were supposed to have a human-sized brain with a simpler cortex, say, less than ten billion neurons packed in there. And indeed, the newborns we checked certainly seemed to fit that model. But there was something in their DNA, something left over from the original sequences we used, that steadily banked that number upward as they matured."

"So they got smarter than they were supposed to."

"They did, indeed. I pointed out that we were in dangerous territory, creating a race fit for labor that might become aware of its own predicament. So I was tasked with inserting an inhibitor sequence in several genes to keep their intelligence level low. But in the end, when we had to leave them, I removed it."

"And that's when they really took off?"

"Yes, it is." Valentine said it with something like amusement. "And already, so fast, they are coming into their own as a freethinking race. They must be treated so. They desire autonomy, and fear human interference."

"Gee, I wonder where they got that?" Gwen said, a sarcastic smile pulling at the corners of her mouth.

Melciéna smiled. "You know me too well, my dear. Toward the end, when I realized the gravity of what we had done—wielding DNA like a sword—I argued to end the program and set the Naiads free, initially with us as mentors, of course, teaching them independence. But the entire thing became moot with the power collapses back home." She looked out the window, into the darkness, with the floodlights beaming through the green waters, the lights flickering and the Naiads swimming by. "Perhaps there is a balance to things. The human race, we humans, we've been pretty full of ourselves for some time, now. Humility has dropped from our lexicon. The Naiads were born of human hubris, designed to serve, not to live. But slavery is never right, is it? And so, the cosmos appears to have deemed it necessary to bring back some balance."

"Are you speaking of Wentaway?" Tony asked.

"You must admit, the big rock flattened the playing field in many ways. Knocked us down a few rungs. And now, our own creation wants to leave Eden and find its way. Do we have any right to deny them that?"

"And so you've become a sort of advisor to the culture that has sprung up around here. Their counselor, their link to the outside worlds."

"Yes," Melciéna admitted. "I've suggested that they be wary of all human contact in light of the Earth's history. Help the people, yes, but do not encourage them to come here."

"Their suspicion seems to go beyond that. The Naiads are vehement about no human contact on their home turf. Why are they so aloof? What have they got to hide? Weapons?"

"There's something down here." Melciéna's eyes glimmered. "All around us. And it's been here for a lot longer than we have."

"Something?" Tony's eyes narrowed.

"Something beyond smoke and mirrors. True magic and wonder." Melciéna's expression melted into one of pure joy.

"Does it have something to do with that flickering light out there? That construction project Thrk mentioned?"

Melciéna waved the question off. "All this is another matter entirely. I'll show you, but not now."

In the corridor, footsteps padded in rapid approach. A young Naiad leaned through the hatch. "Vice Gerent, there is urgent news from the control center."

"Yes?"

"Kyv has been monitoring the frigate; it seems a landing party from the battleship will be deployed imminently."

Melciéna's smile did not falter. "I expected as much. Shall we go?"

"A landing party?" Gwen said. Absentmindedly, she retracted her hand from her diary, which remained perched atop the windowsill.

"Yes," Melciéna said.

"You heard it," Tony gestured toward Melciéna. "Straight from the seahorse's mouth."

Melciéna and Gwen looked at each other, then at Tony. The tension finally split, and they broke into gales of laughter.

Chapter 33
Final Approaches

"Forty-three hours out, sir."

Colonel Carter Rhodes glared out the cockpit window of the *Firecross*, grumbling to himself. "Two days away. Two days. What could they be doing?" He turned to the comms officer. "Peters, can you display that last communiqué again?"

Text scrolled down the screen.

Colonel Rhodes: Your frigate has just dispatched a small craft that seems intent on a landing. On several occasions, we have requested that you not do so, and have made these requests in good faith. The Naiads are adamant about no human contact on their moon. The action of the ship may not be an official one. If this is the case, we have lost confidence in your ability to lead under current conditions. If the action is an officially sanctioned one, we are confused as to your motives, which do not seem to be pursuant to peaceful negotiation. Please clarify immediately, before we are forced to take action against the landing party. Reg Broadwalter, for Melciéna Valentine

"That last line sounds damned ominous. Threatening. Since when did Melciéna Valentine take sides? Since when was she back in the game, even? And what do they think they could possibly use against us?" Rhodes ground his jaw for a moment. "Any further communication from the *Chronos*?"

"No sir. Only static."

"That seems to be the theme around here. Static. It's just not like Chang-Lopes. The guy's strictly by-the-book. And no follow-on transmission from the surface? From the *Belvedere*?"

"None, sir."

"*Damned* ominous. I don't understand why *Chronos* dispatched a landing sortie. Not what I would have recommended, and they know it."

"Something must have happened to change things. Maybe they were threatened?"

Rhodes glowered out the window some more. He nodded, hoping Peters would just leave him alone. Then again, Peters was good to bounce ideas off. Rhodes read the screen again. "If they weren't threatened then, they certainly have been now."

Enceladus was a pinpoint of light to the east of Saturn. Other moons scattered to both sides of the planet. A larger point cast a wary orange glow. Through the cockpit telescope, the golden globe seemed stormier than usual, swirling maelstroms of umbers and tans and, in the southern winter, blues. Storms to match his mood. "Well, nothing for it. When we achieve orbit, we'll need to go to battle stations. Mr. Drew."

"Sir?"

"Prep landing crews from three of our ships. Tomorrow evening I want all weapons at the ready. Anything goes wrong and we'll have time to fix it before we arrive."

"Right away, sir."

"Peters, how's the signal?"

"Still degrading, sir."

"All right, let's tighten up the fleet, get everybody close to each other. We need to be able to talk to one another. Can't risk another loss of comms like we've had with *Chronos*."

***

The arena that served as the center of the Naiad city throbbed with activity. In the blink of an eye, the simple, darkened room that Gwen had first encountered had transformed itself into an active hive of technology. Several large screens had come to life, making their presence known only now. Several Naiads wore earbuds stuffed into the naked holes in the sides of their heads. Others studied the data on the screens. The central monitor, larger than the others, displayed readouts that Gwen recognized as levels of signal strength. But Melciéna was the object of her attention. Everyone in the room seemed to know that she was the one in authority. Even Kyv treated her with supreme respect.

Over the years, it had been so much easier to think of Melciéna as the uncaring villain, leaving the human race behind in its dive toward the Middle Ages. But here in her element, she showed the inner fires of her greatest passion. Gwen could see no evil in it. Yes, her passion had shifted from her own kind to this strange new race, and it was evident that she had refocused her energies some time ago. They could argue where and on whom those energies would be best spent—the point was that she was now receptive to that debate.

Life was nuanced.

But for Gwen, the key mystery of the Naiads remained: how had they come so far, so fast? She stepped to Melciéna's side.

"I don't understand." Gwen swept her hand across technological marvels surrounding them. "How did all *this* happen?"

Melciéna gazed at one of the screens as she spoke. "Yes, my dear. I understand your confusion." She turned to Gwen and gestured to a couple of chairs at the wall. "Come. Sit. I'll tell you a story."

"Vice Gerent," one of the engineers called.

33 Final Approaches

Melciéna held up her hand. "In a moment, Rbq." Turning back to Gwen, she said, "It is important to me that you understand."

They sat. Melciéna met Gwen's gaze with those fiery eyes. "My dear, back then, things got...complicated. Wentaway changed everybody's timeline; it became obvious that the humans were going to be leaving here *en masse*. When I looked at their available resources, rag-tag as they were, I could tell there wouldn't be a whole lot of room for 'the help.' My Naiads were staying behind. So for the last few generations of clones, I wanted to at least give them a fighting chance. You see, contrary to how it sometimes seems on Earth, human intelligence naturally tries to rise to the top. It's part of our nature, our evolution. While I was cloning our Naiads, I was told to inhibit those genetic tendencies. But after Wentaway, I undid that inhibitor. I had no idea how quickly they would flourish. It's a process, of course. They have a short life span, but they talk at 18 months; they're sexually mature at 3. There have been generations of them since you were last here. And I assume you've seen some of what they've done. They're having their own Renaissance down there."

"Yes, I saw the sculpture and the building. But all this, too?" Gwen scanned the room. "It's..."

"It's amazing."

Gwen nodded slowly.

Melciéna stood. "Now, to work," she said, then called across the room, "Do we have a clear channel?"

"Yes, Vice Gerent," a technician said.

Melciéna murmured to herself, "Good job, Thrk." The comment puzzled Gwen, but everybody seemed too busy to ask. Melciéna donned a headset and said, "Open a channel, please."

A Naiad wearing a similar headset declared, "Ship to ground active."

"Thank you, Vrl." Melciéna's tone changed to one of firm authority. "Atlantis West to *Chronos*. Please respond."

"Read you five by four," came a voice in the midst of a fading audio flurry.

"This is Melciéna Valentine. Who am I speaking to?"

"This is Commander Chang-Lopez of the *UFM Chronos*."

"Commander, it appears that you are in process of sending a landing party to the surface. Is this correct?"

"Yes, it is. It is a well-armed expeditionary group."

"Not to be toyed with, I'm sure. And ETA is approximately thirty seven minutes?"

Chang-Lopez didn't answer.

"Commander, the ship is dropping in front of our noses. Everybody concerned can see it."

"Your estimate is approximately correct," he said grudgingly.

"Fine. I would like to request that you have your party remain on their lander. We will come meet you with a small contingent of our experts. We have much to discuss."

"Just who do you represent?" the Commander demanded.

"Colonel Rhodes is familiar with my influence in the outer system. You may check with your superior. If I'm not mistaken, he should be fairly close by now. We will meet you thirty minutes after you make landfall. That is all." Melciéna severed the connection.

Gwen laughed. "Take no prisoners," she chided.

"You bet. We need to keep them on their toes."

"So, I'm curious: how did you clear a channel in all this static?"

In answer, Melciéna turned to a Naiad nearby. "Blq, is it about time for a shift change?"

Blq looked to Kyv for affirmation. He nodded in that awkward, Naiad way. Blq said to Melciéna, "Yes, it is time. I'm sure Thrk is growing tired."

Melciéna turned to Gwen. "We still have some time before Chang-Lopez' crew arrives. It's time you saw the alcove. Tony, why don't you come, too? It will avoid some convoluted explaining on Gwen's part."

Melciéna took the lead, while Blq followed behind Gwen and Tony. Though she knew the general layout of Atlantis West, Gwen quickly became disoriented in the twists and turns of the dim corridors. Finally, Melciéna reached a closed hatch. She put a finger to her lips. "We must be quiet in here."

Blq swung the heavy access open. Beyond it, a small chamber spread before them, a few meters across and deep. A windowed roof arched above their heads, low enough to hurt the unwary. Rivets lined the walls and floor, imbuing the compartment with an industrial flavor. The chamber was clearly not built for comfort or fine decor. Thrk sat near the front pane of the room's wide windows in a central chair, facing out, his eyes closed.

Melciéna gently sealed the hatch behind them. Blq stepped to Thrk's side and put her hand on his shoulder. "Thrk?" she whispered.

Thrk's eyes flew open in surprise. "Yes. Yes, Blq. Is it time already?"

"We have visitors."

Thrk turned and scanned the room as if coming out of a coma. "Oh, hello, Gwen. Tony." He stood. "Vice Gerent."

"Good job, Thrk. Our friends seem particularly in tune today."

"Yes, it is a good day for–" He glanced at Gwen. "—our friends." He then rubbed his temples in a particularly human way.

"Why don't you get some rest?" Melciéna said.

"I'd like to stay, to help. Explain."

"Certainly."

Blq took Thrk's vacated seat and closed her eyes.

Gwen leaned toward the windowsill. Beams of light illuminated the clear seawater. Suddenly, she knew where she was. Off in the expanse, flickering away like an ember in a storm, was that mystery construction site to the northwest. Whatever it was, the glimmering structure had become her North Star. It remained a shimmering spot in the darkness, always there.

Melciéna glanced through the glass, then looked intently at Gwen.

"Gwen, what do you know about microbial radio emissions?"

"Radio emissions?" The question brought her up short, but she searched her memory. "I know we've measured radio waves issuing from *E Coli* bacteria. Even two centuries ago, microbiologists thought that bacteria might communicate amongst themselves using some form of radio signals. We use *Geobacter* in bioreactors, as you do here. These little guys grow electrically conductive nanowire filaments. The microbes attach themselves to an electrode, discard electrons through those nanowires, and we harvest the energy." She paused, a feeling of unease washing across her as she looked out into the sterile waters of Enceladus. "But what does that have to do with us? With this place?"

"I think you have suspected for some time." Melciéna spoke quietly. "All that water out there? All those 'prebiotic' goodies mixed in? The ammonia? The methane and nitrogen? Complex organics? You know what that all points to."

"But we've been monitoring the ice and the geysers for centuries."

"And what have we found? Geysers laden with the fingerprints of life, but no active biology. Directed interference in the radio spectrum that comes and goes, seemingly by design. Is that likely?"

"I never thought so."

"As a biologist, you wouldn't. And you would be right. The global ocean of Enceladus hosts a suite of microbial inhabitants. The closest terrestrial comparison is plankton. It's not just one microscopic creature, but a whole zoo of them, some of them plantlike, others more akin to bacteria. And these things combine on Earth to make a global community across the seas. Here, the Enceladan microbes are a mix, like the plankton in the oceans of the genesis world. And those nanowires and shed electrons you spoke of can be directed. *Are* directed. It's like taking the neurons of a brain and scattering them across the ocean. A global mind. But this zoo makes it almost impossible to figure out the nature of the microbes' awareness, the group consciousness that's going on down here."

"Group consciousness!"

Melciéna took on an indulging voice. "It should not seem so alien to you. Consider the honeybees. Not the micro-drones, but the biological ones they use in the organic greenhouses. Each has a brain weighing less than a milligram, but using their communal intelligence, they coordinate to build hives in precise, hexagonal structures for the entire colony. They live in a multifaceted society with individuals taking on specific roles."

"Drones, workers, the queen–" Gwen said. "I know which role you'd be aiming for."

"You bet, complete with tiara. And think about how they impart data to their comrades, with their pheromones and their complex dances communicating danger, a shift in weather conditions, even location of food, right down to its direction in relation to the sun and the distance to the source."

Gwen shook her head. "But the Enceladus plankton are doing a lot more than locating food. They're transmitting radio waves. As a global community."

A slow grin spread across Melciéna's face. "Incredible, isn't it? And somehow, the Naiads have a link to them. Some more than others. Thrk seems to be the best at it. Blq is a close second. There are others."

With all that had been revealed to them, Gwen was amazed that both she and Tony had not turned to a pile of jelly on the floor. Perhaps the human mind was more lithe than she had thought, more able to take such a tsunami of new and paradoxical inputs. Perhaps.

Tony asked, "And this giant microbial mind has been interfering with human exploration here for over a century?"

Thrk had been listening, but now he chimed in, "It's nothing more than a defense mechanism for them."

"Keep the intruder busy and looking the other way," Melciéna added, "and you can hide nicely in the cocoon of your own sea."

Gwen's heart raced. She felt flushed: A piece of the cosmic puzzle had just dropped into place. "And this, this is why the Naiads have been so averse to contact with the human race. Not just for themselves."

Melciéna nodded. "For those who came before. The Naiads are protecting this world and the indigenous life in it."

"Amazing," Gwen whispered.

Blq sat up and opened her eyes. "The others have arrived. Their ship is waiting a few paces from the main surface airlock."

Gwen gaped at her. Melciéna said, "The microbe hive mind can tell us a lot of things. Let's suit up."

Chapter 34
A Walk on the Sunny Side

TO: Chevalier, Armand
FROM: Rhodes, Carter
RE: PERSONAL TRANSMISSION~SCRAMBLED
TIMESTAMP: 04:32:10GMT

I don't like it, Armand. We've just lost contact with the *Chronos*, like Earth did a week ago, and the fleet can barely transmit amongst ourselves. To make matters worse, *Chronos* has apparently dispatched a landing sortie. This according to radar at our Iapetus base. It seems that whoever is running the show at Enceladus is about to meet up with our military might. Our hand has been forced by events even before we arrive. I only hope Dr. Baré has been able to make some headway in the diplomatic realm, but because communication has been so spotty, I fear the worst. If she has failed, we'll be ready. No matter what, let's hope we'll be heading home soon with those power plants from Thera and Atlantis West. My guess is we'll probably have a few refugees or criminals in tow.

Kyv read the message again. "And this went out this morning?"
Blq nodded. "An hour ago. It will be reaching Earth in a few minutes."
"It is as we had feared. The threat of violence. I feel sorry for the humans. So quarrelsome. So quick to make decisions. Capricious."
"The Vice Gerent was right to be cautious."
"Yes, she was. We must make preparations."

***

Gwen marveled at the Naiads' environment suits. They were bulky, but more streamlined than she expected. Each suit carried not only systems to retain air pressure, but also systems to humidify the interior. Naiads had more environmental needs than humans did. Like the human environment outfits, the Naiads' suits delivered a cocktail of drugs to prevent the bends, though the Naiads needed far less than their human companions.

In all, the vanguard was a party of six. Gwen and Tony walked behind Kyv, who kept Thrk at his side. Two others walked behind the main group, watching. The landing ship stood close to the surface station, towering above them in harsh bastions of metal, shining glass, and jutting weapons. Beyond it to the south, the *Belvedere One* and the *Belvedere Two* sulked on the ice with nowhere to go.

"An impressive show of force," Kyv commented as they reached the lock. Several weapons turrets followed them as they walked. Humans peered out through various portholes. One of them gestured to the group as he opened the outer hatch remotely. They stepped in. There was plenty of room; the airlock was sized to take a large contingent of soldiers.

The crew of the vehicle pumped up the lock and unsealed the inner hatch. The party ducked through the cramped opening and stood before a roomful of armed GIs. One of them stepped forward.

"Welcome aboard the shuttlecraft of the *UFM Chronos*. I am Captain Daniels."

"And I am Kyv. I represent the inhabitants of Atlantis West and the more remote enclaves of the Naiads on Enceladus."

This was news to Gwen. Where were the other Naiads living? They had found Thera to be completely abandoned. How many more were there?

Perhaps these questions were on the Captain's mind as well, but he did not say. Rather, he addressed Kyv, "We believe you have taken several hostile actions against our people. A missile destroyed a scout ship several months ago, and you have been jamming our transmissions to Earth. Please explain."

Gwen watched Kyv. Daniels had tried to put him on the defensive. She wondered if he was psychologically equipped to handle it.

"On the contrary," Kyv said. "The missiles in question came from Iapetus. We monitored their course from here. They were sent from an automated installation set up decades ago. We don't know who first put it there, but it seems to be a simple station, designed to protect this area from unmanned probes. We don't believe it was intended to carry out a hostile act against piloted ships. The loss of your ship was a tragedy. Any loss of life is."

The genesis of the defensive station was no mystery to Gwen. She knew it must have been one of Melciéna's designs to protect the secrets below the ice.

"As for the static in your transmissions," Kyv continued, "you will note that the literature over the past century has documented this phenomenon in detail. Now, we have been wondering why you have made landfall on our world without first asking permission. This, too, would seem like a hostile act."

Daniels froze. His troops looked at him expectantly. Before he could speak, Kyv added, "Whatever the case, there is really no need for us to argue about such petty things. It is likely you were not aware of our presence. We understand this. We have gone to great lengths to keep a low profile here. You are welcome to stay as long as you want. We are happy to bring you some refreshment that we have grown in our greenhouses. However, we must insist that you remain aboard your shuttle and that you stow all weapons. They are not needed here. We have none. You need none."

Gwen had not expected Kyv to volunteer that last bit of information straight to the Captain himself. The man would now know just how vulnerable they were, and they had lost valuable bargaining power. She glanced backward, to where Melciéna stood observing in the corner. Melciéna seemed entirely unfazed, as if she had known Kyv would do exactly this, as if it was all part of the plan. *Smoke and mirrors.*

"I see," the Captain said, not moving.

Kyv glanced at Gwen. Pushing down her own surprise, she addressed the Captain, "Sir, I'm Gwen Baré, and this is my associate, Anthony Vincenzi. Colonel Carter Rhodes has sent us here in hopes that we can negotiate in an orderly way."

"Dr. Baré, the Colonel will be happy to hear that you are safe." He glanced at the rest of the party. "If, in fact, that is the case."

"It is the case."

"Now, to the task at hand," Kyv said. "The Earth is in need of power generators to supplant its dwindling supply. Is this correct?"

"It is."

"We have many bioreactors below, powering our homes, workshops and city. We are willing to share some of these and discuss making more for export."

"I can vouch for the microbial fuel cells," Gwen said. "They have advanced designs that generate more power than anything we have back home. A trade of this technology would be a benefit to the entire global grid."

"I see," Daniels said again, in the same flat tone. Clearly, he was faced with new information, unsure of what to do in the shifting circumstances. Squaring his shoulders, he said, "And therein lies the problem. We have been tasked with removing all the power sources available and getting them back to Earth on a burner, as quickly as possible. The grid is collapsing as we speak."

"Be that as it may," Gwen said calmly, "clearly the situation has changed because of the presence of the Naiad civilization."

"Civilization?" Daniels scoffed, his voice hardening. "We're talking about ten billion people. Earth's entire civilization. *Our* civilization."

"Surely, Captain, there is room for compromise," Gwen said.

"Even if there were, the fact is that there is no time. Our task is to return with reactors immediately. Until I hear other orders, we will be proceeding accordingly, even if it means the use of force."

Kyv spoke reasonably. "You may certainly make that choice. It is the least preferable for all parties. Until we have assurances that your weapons will not be used, you will not be returning to your ship. We do have the capability to incapacitate not only your lander but also your battleship."

"Incapacitate?" Daniels bellowed.

"I am not suggesting this as a course of action. I am merely informing you of some... *some* of our capabilities."

Gwen stepped in, "It is imperative that we come to a peaceful understanding. As you say, time is of the essence, so it's now or never."

"I will need to consult with my superiors."

"Your other ships will be here before local nightfall," Kyv said. Daniels shifted, unnerved by the fact that Kyv already knew about the approaching armada. "We shall keep communication open on the specified channel. We will leave you to *consult*." Gwen could almost hear a smile in Kyv's tone.

The group moved toward the lock. At the back, still in her helmet, Melciéna observed, anonymously.

***

TO: Rh-des, Ca--er
FR-M: Chev--ier, Arm-nd
RE: PERS--AL TR---MIS------BLED
TIM--TAMP: 06:53:07G-T
Col--el Rhodes, it is good th-t forces are ar--ving now. It --pears that we've just lost the entire easter- seaboard o---frica and the sub-cont--ent of India. They're saying Europe shou-- fall within th-----hina and Japan will be hot on their heels. The Antarctic Commons, Australi-----ew Zealand may be okay for now. They say if we don't get a twent----rcent power surge in the next sev-------ks, damage may be irreparable and we'll have to start ---r on vast tracts of grid. Sorry for the bad news.

So this is it, Rhodes thought. They were now out of time for negotiations of any kind. And the message had barely gotten through, so he was on his own out here, with no promise of aid from Earth or elsewhere. It was time to take whatever was available by whatever means necessary. He mentally braced himself for what might well be an all-out battle.

Rhodes called to the ship on his port bow. "Sanders, any news from the surface? From Daniels?"

"None, sir."

"And Chang-Lopez on the Chronos?"

"Same story, sir."

Rhodes sighed and gritted his teeth. Through a clenched jaw, he said, "Peters, keep at it. We need to get through. For now, get me the fleet-wide frequency."

"On line, sir."

Rhodes keyed his mic. "This is the *Firecross*. Go to full battle stations."

***

Kyv had already gone to the control center by the time Gwen pulled her helmet off. She leaned over to Melciéna and said, "Can the Enceladan plankton really do all that?"

"Truthfully, I'm not sure. They can disrupt electronics, confuse navigation, maybe even command a ship to descend to the surface of Enceladus. But we've never asked them to do anything on such a grand scale before. On the other hand, they have the power of an entire global ocean to project."

It was then that Gwen noticed the darkness in the porthole behind Melciéna. Out in the Enceladan sea, something had changed. Gwen's North Star—that flickering, ever-present light—was gone.

She pointed toward the emptiness. "Melciéna, what was that out there? What kind of construction?"

"Just a little surprise. A revolutionary one, actually. You will see soon enough. It's a ship we've been working on." She waved her hand, as if swatting away an annoying fly. "Now, you and Kyv should have a chat about things. Strategies. I will leave you to it. I have some business to attend to." She turned to leave, but hesitated. "It is so good to see you again, Gwendolyn. I'll come see you at Venus. Or perhaps in Philadelphia." She surprised Gwen by wrapping her into a tight hug. It was an embrace that made up for years of lost time, an embrace that reminded Gwen of her mother. She held on for a moment.

"Philadelphia?" Gwen mumbled into her shoulder. Did Melciéna know something she didn't? If history was any lesson, her enigmatic aunt knew far more than she let on.

"We both have things to ponder, do we not?" With a flourish, she snapped her fingers and handed Gwen a tube of sunflower seeds, seemingly out of thin air. Then Melciéna Valentine turned, stepped into the corridor, and left.

Gwen raised her hand to brush a lock from her forehead and felt something scrape against her wrist. Inside her sleeve was a playing card: the Queen of Hearts. On the back of the card, someone had scrawled the words, *SPEED IS EVERYTHING*. Gwen sprang to the hatch to call after Melciéna, but she had already vanished.

***

As Gwen changed into fresh clothes, the comms set in her helmet beeped. It was the first sign of life the headset had given her since she had arrived in the heart of Atlantis West. Finally, the *Belvedere* was getting through again. She picked it up and held it to her ear. The scratchy voice on the other end was saying, "How read? *Firecross* to Baré. How read?"

Gwen keyed her mic. "I read you, three by five. Can you boost your signal?"

The voice on the other end changed pitch. "Gwen, it's Rhodes. My comms guy just busted through the static, but I'm not sure how much time we've got. I need you to take immediate action. Work around the Naiads if you must, but by all means, disconnect the reactors for transport."

"Disconnect all the reactors?" she said.

"Correct," his voice crackled through the hiss.

"But that will kill every living thing in Atlantis West."

"No worries, we'll get you and Tony out in plenty of time. We've got our landing party there, as well as Broadwalter's yacht."

Rhodes had some catching up to do. Gwen could see clearly now what he was, and what she had been not so long ago: so preoccupied, so laser-focused on one way of seeing something that life had become a caricature around it. She had to do something about it.

"Colonel Rhodes, we need to meet. There are greater things at stake here, and solutions that might benefit all parties."

She heard the blizzard of radio noise.

"Colonel Rhodes?" Static. "*Firecross*, how read?" They were gone.

Gwen rushed into the corridor and headed in the direction of Kyv's control center. The corridors all looked the same, and she made several wrong turns before she got her bearings. Finally, she came to the great door. It blossomed open to a room bustling with activity. Melciéna was not there. Thrk stood next to Kyv, who was seated at his dais in the center.

Kyv stood when he saw her. "Dr. Baré, how may we help you?"

She had moved so quickly, in a panic, that she had not taken the time to evaluate the possibilities. Rhodes was headed here for confrontation. He'd be here soon, bringing to bear a whole lot of firepower. Time was short, but what to do with it? Was it wise to tell the Naiads, or should she let things unfold? Was she observer? Go-between? Champion of the underdogs? Who were the underdogs here? While Rhodes had the distinct military advantage, he came from a hurting world. If she helped the Naiads, she was committing to one side, to one paradigm, was she not?

Kyv and Thrk stared at her, looking almost frog-like. *Time to decide.*

"Kyv, Rhodes just communicated with me, through my helmet headset. He has commanded me to disengage all the power plants here."

"That would have severe consequences."

"I'm thinking it's time to negotiate."

"As you know, we are open the possibility."

"We would need the help of the global plankton. I need to be able to talk to him before he mounts some kind of major weaponized–" She paused. Attack? Maneuver? *"Thing."*

Kyv turned to his aide. "Thrk."

Thrk left the room, undoubtedly headed for the alcove.

"Come, Dr. Baré. Sit at my console. We must give Thrk a few minutes, and you and I must consult."

The two spoke in hushed tones, a confidential conversation in the midst of the busy room. Gwen nodded. Kyv rocked his shoulders.

Kyv addressed one of the Naiad technicians. "Prn, please open the frequency of the Earth crews so that we may monitor."

A speaker next to Gwen's console brought a hail of static. Kyv turned the volume down slightly, and sat. Gwen listened to the haunting sound. The audio maelstrom carried the voices of a thousand banshees, of the ghosts of the dead, of the ammonia snowstorms of Saturn and the methane monsoons of Titan.

And suddenly, it was gone. Silence.

"I believe you have a clear channel," Kyv said. "Prn, patch us in to the *Firecross*."

"Baré to *Firecross*. How read?"

The voice on the other end came in clearly. "*Firecross* to Baré. Five by—yes sir, sorry sir."

Gwen could hear some shuffling on the other end, and then the voice of Carter Rhodes. "Gwen! Rhodes here. Are you okay?"

"Hi, Carter. Yes, I'm fine. I'm reeling a bit from a few new revelations here."

"Yes, I understand you have lots of company down there."

Rhodes may not have known her circumstances, so she clued him in. "Right, and they are with me now. In fact, you told me to be a diplomat if needed, so I've been working with the Naiad leader, Kyv."

"The Naiad...*leader*?" Rhodes sounded astonished.

"I think Kyv and I have come up with a plan that will make everybody happy."

The static remained calm, the transmissions clear.

"Kyv. And you. I'm open to just about anything, Gwen."

"Good. Can we have visual?"

"I'm afraid with a military ship like this, protocols dictate that we remain voice only."

Gwen shook her head. This was not the most encouraging start. "Read you. Our proposal involves some new technology down here. They've developed bioreactors far more powerful than anything we've got, and they don't require the years of seasoning to get them up and running."

"Sounds promising."

"It is. And the Naiads have the capacity to mass-produce them for Earth. They can begin almost immediately."

Rhodes did not respond. The channel was clear. The interference had abated. He simply wasn't talking. After an agonizing moment of silence, he said, "I'm assuming they have some conditions, something they want in return?"

"Yes, the Naiads have some long-term plans that will affect this, but they're willing to emplace energy sources as soon as you agree. For the long term, they'll need more raw materials to make the stuff. The primary condition is that there be no human physical contact with Enceladus. No landings. That's a given for them. Prime directive. Any necessary meetings and exchanges are to take place in orbit, or even at a neutral location like Iapetus or Ganymede."

"We'll need to discuss this on our side, of course," he said.

"Sure. Of course. Details need to be ironed out. But in the meantime, Kyv has offered to ship several reactors to Ganymede for the Earth technicians to examine. I think you've already got people standing by there, right?"

"Yes, we do. We have some technicians on board here, and one stationed on Iapetus, but they aren't as qualified. I've got a Ganymede team that could examine them. But I'm worried about the timeline. The grid has essentially collapsed. Even if we could plug a thousand of these in right now, I'm not sure we'd be in time. In a fast ship, it's weeks to Ganymede and months more to Earth by heavy transport."

Gwen knew the power curves of the world's network, and knew how quickly they would respond to an infusion of supplemental energy. Rhodes' suspicion of the proposed schedule seemed unreasonable. He was delaying, remaining noncommittal.

"What if we ship to Iapetus and broadcast images to the Ganymede team? Or even transmit from here directly?" she offered.

"No good. We need to measure output in person, investigate the guts, look under the hood, so to speak."

"Colonel Rhodes, next to me is Kyv, the leader of the Naiads here. He would like to talk to you."

Kyv entered the conversation. "Colonel Rhodes. We are prepared to begin this task immediately. It appears that our technology can help the people of terra, and we ask little in return: for us to maintain our independence, and possibly to establish trading networks, which we can work out once your crisis is stabilized."

Gwen thought Kyv was being quite generous. And trusting. She wasn't sure Rhodes had earned this trust, but they had to put all their cards on the table. She said, "Time precludes our negotiating in full at this moment. We have a crisis to solve, and it appears to me that the Naiads offer us the solution."

"How much power output would be available to us, initially?" Rhodes asked.

Kyv nodded to Blq, who keyboarded some numbers. "The particulars are on their way to you. We will send you what we can spare now, and begin fabricating more for you immediately. The new bioreactors can be in your hands by next spring, northern hemisphere time."

This time, the pause was long enough that Gwen began to get nervous. The room seemed too small and, for the first time since she had arrived, too warm. After an eternity, Rhodes' voice came over the speaker. "That's a long time to wait, but I accept your terms with thanks."

Rhodes signed off. Gwen felt the burden of her entire mission dissolve. This was it: a final solution to the Earth's crisis, and assurance of the Naiads' freedom from outside forces. She wished Melciéna had been here to see it, but she had done what she was so good at doing: disappearing at a critical moment. The next several years on Earth would be difficult, and technologically tricky, but the home world could finally see the light at the end of a very long and dark tunnel. And with a diplomatic agreement between the Naiads and the humans, Gwen could finally head for home, where there would be ample supplies to restock her dwindling sunflower seeds.

Gwen smiled at Kyv expectantly. She was ready for a celebration. But Kyv said, "If I was your Colonel Rhodes, I might be skeptical of the Naiads' trustworthiness."

Chapter 35
The Dark Side

Ganymede was preparing for an eventual shipment of two advanced microbial fuel cells for inspection. Peters watched Rhodes as he drifted anxiously from one window to another.

"Sounds like we have reason to celebrate, don't we, sir?"

"Caution is the order of the day, Peters. Caution. We don't know them. I don't trust them. How do we know that Gwen wasn't being coerced? How do we know what hidden agenda the Naiads might have? I can't believe I'm talking about Naiads *even having* an agenda. I wonder if someone else is down there, pulling the strings. Refugees from Miranda. Martian outcasts. Who knows? Last I checked, Naiads didn't have the capacity for agendas or trade agreements. No, there is more here than meets the eye, I'm afraid."

Peters paused before speaking. "Orders, sir?"

"Have the fleet remain on battle stations. Ganymede is too far away. We're going in."

"To take their reactors?"

"That's why we're here, isn't it? There really is no other option, sadly. We've got to take Enceladus' generators directly from the ice moon with our own ships. Peters, this is not a belligerent action. This is a heroic one. Earth is counting on us. Besides: we are simply coming to reclaim what we bestowed upon this planet in the first place. Nothing more, nothing less." The bluster in Rhodes voice was tentative. Peters looked uncertain. "Well?"

"Sir, if we take their power plants, I assume the entire settlement will shut down."

"Yes, yes it will."

"And all those green guys?"

"Of course, I would never propose killing a bunch of…of Naiads. It would be a waste. We'll bring transports. From somewhere. Relocate them to the oceans of Earth."

"Could they even take the heat? Or the gravity?"

"Maybe we resettle them at Europa. There's still a bit of operational infrastructure."

"Yes sir." Peters sounded unconvinced.

"Reservations, Peters?"

"Colonel, they'll all die as soon as we pull the plug. We have no transports near enough to relocate them in time."

"Could be a problem. Yes. Have someone look into what's available. But first priority is to get that energy back to Earth. Nuclear, microbial, candlesticks if they've got them."

He gazed at the glistening landscape curving below them, and listened to the relentless radio blizzard of Enceladus.

***

Gwen was organizing her room, getting ready to pack. If all went well, she would be on the first flight to Ganymede with Carter Rhodes, and then home. Captain Daniels and his crew had not ascended to the comfort of their ship in orbit, but rather were standing by on the surface, waiting for something not clear to Gwen. His well-armed ship kept Gwen on edge, but she felt too triumphant to be overly worried. Taina remained at the ready to take Gwen to the *Firecross* when it was ready to head for the Jupiter system.

A knock on her door interrupted her joyful preparations. She opened the hatch to a worried Tony.

"Come on in. What's up?"

He stood motionless in the doorway. "Mi amica, you have to come see this."

Gwen followed him down the corridor to a lift and down to the main concourse. They stood at the opening of the expansive spine of Atlantis West.

Tony murmured, "The Naiads, they are on the move."

Beneath the soaring arches and amongst the shining sculptures, long trains of the green creatures seemed to be exiting the southern part of the settlement, moving north into the causeways and halls of the other side. Through the clear floor, Gwen and Tony could see lights migrating northward.

"Maybe they're involved in the preparation of those reactors going to Ganymede?" Gwen knew it was a weak guess at best.

"Yes, all two of them," he scoffed.

"Let's go back to my room. You and I need to investigate."

"Maybe this is some Naiad festival or something," Tony said halfheartedly.

"Sure," Gwen said. "Festival of lights." But once in her quarters, her window told her a different story. Across the wingspan of Atlantis' butterfly, the outer lights on the south wing had all gone dark.

Tony pressed his nose against the glass port. "What could the Naiads be up to? It's like they're shutting down the power over there, closing down half of the entire settlement."

Gwen froze. "Yes. Yes, that's exactly what it's like." She joined him.

While Tony was studying the darkening of the southern sectors, Gwen noticed that something was happening to the west. At the spot where her North Star had been, another glow began to burn, but it was quite different from the starlike object that had vanished. This glow was amorphous, purple, and directional. Although it illuminated a large area, the center resolved itself into a cone spreading downward. The soft pyramid of light slowly rose, eventually dimming away into the darkness above.

Chapter 36
Goings and Comings

Taina continued to monitor all activity aboard the warcraft that lingered near the access to Atlantis West. For a shuttlecraft, it was a behemoth. It lounged on the ice like a beached beluga with talons, and those talons were sharp: laser cannons, electronic weapons, and an assortment of missiles primed to assail any unsuspecting prey. She detected periodic radio transmissions from the ship, which had increased once the background static had died down. Now, the ship seemed to be checking in regularly with those in orbit overhead. Taina listened in on their transmissions, mostly routine and boring communications. Then came another missive, this one directed at her. It was not from the shuttlecraft. It was from below.

Once she had received the message, she called toward the stern of the *Belvedere One*. "Jerome, it's show time."

The big man came forward and peered into the cockpit. "Should we go outside?"

She jutted her chin toward the hulking shuttle. "We don't want to alarm our friends next door. She said to just keep an eye out to the south."

Jerome popped open a tube of chips and they sat side-by-side, snacking, watching, waiting. Before they saw anything, they felt it: a deep rumbling in the ice that shook the floor of the yacht. Beyond a row of low-lying hills on the horizon, an impossibly bright glowing object broke through a titanic cloud of water vapor. It quickly rose, accelerating into a streak of indigo light and arcing in a direction opposite the military traffic above. The southern sky glowed with a ghostly blue, the telltale sign of cryovolcanic activity in the Tiger Stripes beyond the horizon. The glowing point of light faded away above the curtains of erupting geysers, doglegging in a seemingly impossible maneuver before disappearing completely in the starry firmament.

"Godspeed, Melciéna," Taina whispered. She reached over to Jerome. "Got another chip?"

***

Gwen and Tony stepped into the control center. Kyv stood at its heart, beside Blq and Thrk. When Kyv saw them, he motioned them over, then pointed to a central screen. "It appears that there are five ships, all now in Enceladus orbit. These are in addition to the one that settled into orbit from Iapetus earlier."

"That's a lot of firepower," Tony said.

"Yes, it is. I'm afraid it is a symptom of a low level of trust on the part of Colonel Rhodes. Sadly, we must take other steps to secure the peace."

"Other steps?" Gwen asked.

Kyv nodded in the Naiad way. Gwen's mind kept going back to the lights of southern Atlantis West blinking out. Was the dimming of Atlantean lights one of the steps Kyv was referring to?

"Kyv, what's with all the darkness? Have we had power failures?"

The Naiad kept his eyes on the screen, but said, "We needed to repurpose some of our reactors." He gazed at the screen intently, and she sensed she wouldn't get any more information out of him. Not for now.

Gwen studied the elongated blips on the screen. Each had a block of data associated with it. The ships were of different sizes, with Carter Rhodes' frigate in the lead. It was the largest of the cruisers.

"Here they come," a Naiad tech called out. Next to the green blobs, white pinpoints appeared, blinking like Christmas lights as they separated from their mother ships. On the larger ship indicators, more than one pinpoint flashed to life. They were all dropping downward, all headed toward Atlantis West.

The white noise rose in the speakers again. Pops and hissing filled the room.

Kyv barked at Thrk, "We must keep the interference down. The channel must remain open to the Earth."

Thrk hurried from the room.

"Why do you want to give Carter's ships clear reception?" Gwen asked.

"So that they can receive a very important message."

"From you?"

"From Earth. It should be coming in just a few minutes."

The static died down abruptly.

A nearby technician said, "Kyv, ships are ten units in altitude, seventy downrange. ETA in twelve minutes."

Thrk entered the room again. Kyv told him, "We must go and greet them. Suits, everyone."

"May we?" Gwen said, gesturing to herself and Tony.

"Please. It is only proper. But I would request that you do something first. If you would, please inform their leaders that we have more advanced technology down here than they expect. In this way, they will perhaps be more cautious in their approach, and reasoned in their actions."

Why was Kyv willing to risk a meeting on the surface? There was no time to ask. Gwen hurried to the comms center with Tony, transmitted a brief and somewhat cryptic message, and dashed to get her suit.

**

Kyv, Thrk, Gwen and Tony stood outside, four tiny figures against the backdrop of white, blue and gray. They stood a distance away from the entrance, its airlock and domes, facing the velvet sky. Saturn was nearly full, and the sun almost set on the opposite horizon. Above, among the stars, the flash of a dozen lights announced the arrival of Colonel Carter Rhodes and his military forces.

Rhodes' advance ship pirouetted overhead, flickers jetting from its belly, weapons protruding like the quills of a porcupine. In their headsets, the four could hear the Naiads in the control center. "Lead ship at thirty meters...twenty-five..."

The ship began to kick up ice crystals from the surface. Gwen found herself in a wild blizzard, the ship and airlock and ground and sky and her companions all blending into a dull gray fog. Her senses overloaded in the fury of the landing craft, and she felt dizzy. *Keep it together: this would not be a good time to faint.* Momentarily, the maelstrom cleared. The ship rested on the ground, towering before them, an impenetrable wall of metal and weaponry.

Kyv radioed the Naiad control center. "Any word?"

"None," came the reply. Gwen could sense a change in Kyv's posture, a tightening.

The ship's belly hatch swung open and ramped to the ground. In a flash, two dozen armed Marines ran out and stationed themselves on the ground, weapons raised. Two others came through. One wore the suit of a Colonel. Carter Rhodes.

The military personnel stood their ground. Kyv, Thrk, Gwen and Tony formed a stationary front between them and the access port to Atlantis West. All those guns, and all the gadgets Gwen couldn't identify, made her nervous. With some fumbling, she tapped her helmet earbud and stepped forward.

"Hello, Carter."

"Gwen! I'm glad you're part of this party. Tell them we mean them no harm, and that we will leave them alone as soon as we take the reactors."

"Colonel, you have not thought your actions through." The voice was Kyv's, and at first Rhodes seemed completely confused as to the source. Kyv continued, "What you propose would be a detriment to the entire colony. It is inhabited, as you now know."

Rhodes took a defiant step forward. "It seems to me that you've been hiding more than you've let on. Just how did you get that ship headed to Earth? Where did you get the plans for an interplanetary cruiser? Who designed it for you?"

Before he could answer, a message came through for Carter. Broadcast on the common channel, it resonated in the helmets of everyone present.

"Colonel Rhodes, urgent message from Earth."

"At this moment? Are you paying attention to what's going on out here?"

Rhodes could have stepped away, but the move would have provided no privacy. He could have raced back to the cockpit, but it would have taken him ten minutes to cycle back through the lock. With an edge of irritation, he said, "What is it, Peters?"

"Sir, the Naiads have landed on Earth."

"Now? Already? What is it, an invasion of some sort?"

"No sir. They showed up with two dozen microbial fuel cells."

"Operational reactors?"

"I'll say. And they set to work right away, teaming with ground crews. They sent advanced word before they arrived. And the Naiads have brought far more than they agreed to. These reactors are more efficient than anything our experts have seen."

Rhodes walked over to Gwen, and turned to the two Naiads by her side. "But how? How did they get there so fast? Nobody can travel that fast."

"Actually," Kyv said, "we can."

Chapter 37
An FTL on the Wall

A stunned Carter Rhodes invited Gwen and the others to accompany him back into the ship for refreshments while they awaited further word from Earth. The Marines remained at the ready outside, weapons trained on the domed station.

They sat down in the small ship's galley. Rhodes looked at Kyv, puzzled, then smiled at Gwen as if they were long-lost friends. She supposed that, on some level, they were. They did have some history, and had built some trust. But there was a more burning issue, a pressing question that Gwen had for Rhodes. "I don't suppose you guys have any sunflower seeds on this tub? I'm about out, and I get really grumpy when I'm out of my sunflower seeds."

"She does," Tony confirmed.

Rhodes ignored her request for the moment. "So, Kyv: how did you do it?"

"I prefer that we wait for word from Earth before answering. We want you to be assured that our reactors are working as they should."

Gwen could tell Kyv needed to buy some time. She tapped the table. "Carter, sunflower seeds?"

Sighing, he said, "Salted or plain?"

Peters appeared at the hatch. "We have word from Headquarters, sir. The Naiad power plants are working at full capacity." He held up a tablet and began to read. "'Helping to staunch the loss of power in parts of Africa and southern Europe.' The European ones just came on line, apparently. They ask if and when they might expect more."

Rhodes looked at Kyv. Kyv looked at Gwen. Gwen was rolling her eyes about the sunflower seeds. Would she ever get them?

She cleared her throat. "Now that you have a substantial set of power plants, we have something further to discuss."

"You've got my attention. How did you pull it off?"

"Pull it off?" Gwen asked.

"The trick with the bioreactors. Did they already have them inside the orbit of Mars when all this started? Did you know how this would play out all along?"

"The Naiads have requested that we discuss nothing until you have all your troops stand down and safe your ships' weapons systems. Just until you and we feel we have come to an agreement," Gwen said. Rhodes was silent. "Do this in good faith. The Naiads have acted accordingly."

"That's a tall order for a group of Marine vessels."

"Not so tall compared to saving a global power grid, wouldn't you agree?"

Gwen could see the struggle on Rhodes' face. If the Naiads were deceiving them, if they were at the ready with weapons of their own, Rhodes' forces would be sitting ducks. But he now had the evidence of the remarkable Naiad energy systems back home. He too would have to exercise some faith—not blind faith, but informed by the strange events he'd already seen. "Fair enough."

Within minutes, the dreadnoughts had stowed their massive weapons, and the Marines came inside, looking bewildered. Of the three large landing craft, Rhodes sent two back to orbit. The only military presence on the surface now was his ship and the *Chronos*' smaller one.

After a flurry of activity, of barking orders and dashing from one vehicle to another, Carter Rhodes sat down before the Naiad entourage once more. "Satisfied?"

Gwen put her hands on the table, interlacing her fingers. "Thank you." She leaned toward him and lowered her voice. "Carter, the Naiads have a new vision for the solar system."

"Well, at least they're not overextending themselves. *The solar system?*"

"Something significant has happened here, something that goes far beyond the energy needs of good ol' terra firma. And you sense it, I can tell. You see, while we were busy pulling back humanity from all corners of the system, hunkering down and just trying to survive at home, Melciéna Valentine never stopped being an engineer. But she was spending less time on genetic engineering and more on propulsion. Those reactors that arrived on Earth, the ones cranking out energy right now, were powering Atlantis West just two days ago."

A look of slow realization crossed Rhodes' face. "FTL?"

"Not quite," Thrk interrupted. "Dr. Valentine and our engineers haven't broken the speed of light, but we have come close. Our new transports travel at about eight tenths of the speed of light. Our new vehicles can make it from here to Jupiter in half an hour."

Rhodes shot to his feet, wavering briefly in the low gravity. He turned right, then left, looking out the window and back in again. "This—this will change everything. Everything!"

Kyv spoke. "It certainly could, and that is what we are concerned about. This technology is a powerful tool. Tools can be useful in good contexts and dark ones. Consequently, we will be the bearers of this technology. Only the Naiads."

Rhodes was a quick study. He locked eyes with Kyv, then with Gwen. "I see. What do you propose?"

"Carter, you're familiar with maritime history. I could tell by the décor in your office. One of those beautiful models you had was a Minoan ship."

"Yes, the Minoans. One of my favorites."

Tony had apparently been content to watch the remarkable proceedings unfold, but now he put in, "I'm sorry. I know about these ancient Aegeans, of course, but what is the connection?"

Gwen's eyes were charged with energy, an energy that infused her voice as well. "Three thousand years ago, the Minoans had the most advanced navy in the world. They ruled the Mediterranean, not by force, but by trade. They had the best ocean-going ships. The ancient Egyptians had the most powerful armies, and they spoke of all the other people in disparaging terms, *except* the Minoans."

"They called them the Sea Peoples." Kyv said.

All eyes were on Gwen as she continued. "The Minoans tied together the entire known world at the time with their advanced technology and diplomacy and trade."

"The Naiads," Rhodes intoned. "The new Sea Peoples."

Gwen grinned. "That's it. The Naiads are offering to become the guardians of near-luminal travel. With it, they will bring prosperity to the entire system. Valentine's NLT technologies will shift the balance of interplanetary civilization. With them, they can tie together the Earth/Mars system, and they can open up the outer system as never before. They had those reactors to Earth in less than two hours. How many weeks did it take for you to get here, and at what cost?"

"Near luminal travel," Rhodes murmured. "NLT? I like it."

"I prefer a good BLT," Tony said, looking surprised to find that he'd spoken the thought aloud.

"We will supply more of the much-needed bio-energy generators to Earth," Kyv added. "And we will supply transport throughout the planetary system. But the arrangement does need to have trade-offs."

"Ah, the catch," Rhodes said. "Trade-offs such as?"

Kyv took the lead. "Enceladus is poor in construction materials and other resources, so we will set up trade in those areas."

"Above my pay grade," Rhodes said. "That's for the diplomats to work out."

"But you do have influence," Gwen said. "After all, you are saving the world."

Kyv held up his hand. "But there is one more thing, and it is the most important thing of all to us." The only sound in the room was the periodic blip from a screen and the quiet environment fans. "Enceladus must be off limits to direct physical contact with the humans or their robots. This is a non-negotiable condition set down by the Naiads."

"It seems a very territorial edict," Rhodes said. "Wouldn't things be more efficient if Earth set up a port here?"

Kyv said, "You are thinking in terms of life before NLT. Sub-light travel will change the dynamic. There will be no need of a port. As I said, this is a non-negotiable term. You must decide between Enceladan sovereignty and prosperity for all, or our enslavement and perpetual Earth darkness. It is a simple decision, really."

"A or B," Gwen added. "And I do suppose that's also in the diplomatic realm, but again, you will have weight there. Carter, we're counting on you to work this out."

"Gwen, I–" Rhodes played with a pen, put it on the table, picked it up again. "It's a lot to get one's head around."

"Isn't it? But isn't it great?" Her eyes widened.

"It is. Of course, it is. The possibilities are…yes, it's great. Amazing."

Chapter 38
Long Distance

In the coming days, Rhodes and his crew collaborated with the Naiads on logistics for the coming trade agreements. A diplomatic team from Earth became the first humans—besides Melciéna Valentine—to travel at near-light speed. They arrived to a warm welcome at the Enceladus surface station, and treated to a virtual tour of the submarine world of Atlantis West. The activities had been helped along in no small measure by a surprising lack of radio interference coming from Enceladus.

Tony peered through the door of Gwen's quarters. "Mi amica, are you packed for the journey home?"

"Just about. You ready?"

"More than ready. I want to see some Mediterranean palm trees as soon as possible. This place is a little cold and dark for a good Italian like me."

"Oh, I don't know; the place is sorta growing on me."

Gwen's personal set pinged. The voice was Thrk's. "Incoming message for you, Gwen."

"Where's it from?"

"Iapetus. It is the Vice Gerent."

Tony stepped to the hatch. "I will leave you to it. Tell her ciao for me."

Gwen took the call. "Melciéna!"

Melciéna's image snowed into focus on Gwen's personal screen. "Hello, my dear. What a fine job you have done." The one-way radio delay was twelve seconds, so Gwen had to wait between bits of conversation.

"*We* have done a fine job. Big effort with lots of moving parts."

"Yes, yes, these things take a team. I must say, I am beginning to like Colonel Rhodes. Reminds me of your father, in some ways. He has a good head on his shoulders. He seems quite reasonable. I think you should give him a virtual tour before he heads for home. Don't you think so?"

"I suppose that's up to the Naiads. A tour might take some of the mystery out of things so it's easier for them to give the Naiads their privacy."

Melciéna bobbed her chin slightly. "Yes, it was Thrk's idea. I think it's a good one. Might diffuse some of their curiosity. Just don't show him the alcove. Some things are best left concealed."

"What's going to happen with the indigenous life here?"

"What needs to happen. The Enceladus plankton must be left on its own, left to thrive in the primordial seas on this remarkable ice moon. The Naiads have become the guardians of the global life within the oceans of Enceladus. We must see that it stays that way. Perhaps one day, the Enceladans will desire more contact with the outside worlds, and the Naiads can be the link. The Naiads have moved on from being servants of humanity. One day, they will be diplomats for the Enceladans. But beyond that, the servants of humanity have become, in a sense, their benevolent overseers. The Naiad status as traders of the solar system will undoubtedly become preeminent. In time, humanity will look to them as superiors in trade and transport. That is a good thing. Now, I must go. Places to go. Be well, my dear. I hope a woman of her thirties doesn't mind this centenarian telling her how proud I am of her. I hope to see you soon. You made all this work. Kept a lid on things when others wanted to use more exciting modes of operation." Melciéna paused, looking off screen, and then came back again. "Your parents would be as proud of you as I am."

Gwen appreciated the compliment, but for some reason, she realized she didn't need it. Her parents had been on their own journey, and had lived life well. She had embarked upon something quite different, finding her own way. Living life—with all its love and betrayal, sacrifice and cruelty and heroism, its tragedies and glories, magic and wonder—and living it well, was something she was determined to do.

"Oh, and have a good trip back," Melciéna added with a shrewd expression before disappearing from the monitor.

Gwen stared at the empty screen. Her unit pinged again, this time from Tony. "Gwen, Kyv wants to see us off. We're in the control center."

She arrived to a scene of commotion. The control room was alive with activity. Kyv, Thrk and Rhodes were all at the center dais. Gwen leaned towards Tony. "This is the first time I've seen Naiads actually bustle."

"Who knew they could?" he said.

"It seems the time has come for the humans to depart," Kyv called. "Colonel Rhodes was telling me that his view of the Naiads has changed a bit in the last few days."

"I would hope so," Gwen shot back with a smile.

Rhodes grinned. "I've had about thirty more revelations than I can handle. I think I'm just beginning to see the possibilities. With that new drive of yours, there are no launch windows. No waiting for that planetary dance to unfold. You just hop in and hit the gas." He reached out and shook Gwen's hand. "Of course, that's just the hardware side of things. Not to mention the biological, the philosophical. So much to ponder. So many textbooks to rewrite. Thank you, Gwen."

She tipped her head to one side, not sure of what he meant. Was he leaving without her?

"Thanks for all your hard work on this. For playing the energy expert and the diplomat. And for having the wisdom to know when to play each role."

He reached out and shook Tony's hand. "Fine job, Dr. Vincenzi. As we agreed, I owe you an espresso in Milan. Or maybe Padua."

"Yes," Tony said, "somewhere they know how to make them."

"I look forward to our continued relationship," Kyv said, shaking the Colonel's hand.

Rhodes waved to Gwen. "See you back at home. Daniels has already taken his crew back up, and the rest of the fleet stands ready for Earth departure. When you return, we can debrief."

Rhodes and his small contingent left for the elevators and the shuttle home.

Gwen said, "You certainly got the Colonel's attention, Kyv. And mine, too." She closed her eyes and took in a deep breath. "There's nothing like gaining an appreciation of life a little."

"Even life that is very, very small." Kyv held up a finger, its web pulling on the others, and left Gwen with a bit of oceanic wisdom. "From one of the old literature files from Mars, or perhaps from Earth times, I memorized a quote. The author was talking about life across the cosmos, and he said, 'Treat it well. Beneath the Shape, you share the Godhead.'"

"Good words," Gwen said.

Kyv nodded at Gwen, and at Thrk. "And as a very wise woman once assured me, we—all of us—will make those words live. Now, if the two of you are ready, Taina is waiting to take you to the ship that will carry you on the short ride back to Earth."

Gwen gave him a sideways look. "A short ride?"

"We have arranged transportation for you to the hub at terra, on one of our new ships. You should beat Carter Rhodes home by weeks."

Chapter 39
Taina

The *Belvedere One* lifted off of the frosty plains of Sarandib Planitia at high noon. Gwen sat in the cockpit beside Captain Taina Maes. The pilot had dropped her mysterious façade. She had spoken to Gwen like an equal ever since she returned from the depths of Enceladus. Maybe Melciéna had told her to give Gwen the red carpet treatment.

"*Belvedere One* approaching rendezvous orbit," Taina called into her headset. She tapped several buttons on her screen, took off the headset, and sat back. "There. We should dock in an hour or so. I've got her on a slow burn so we have some semblance of gravity. Tea?"

"Sure." Gwen followed Taina aft. As Taina punched in commands on the beverage station, Gwen looked toward the rear. The door to the captain's quarters was ajar. A portrait of some old-fashioned queen or princess or duchess stared out from Taina's monitor. It was a facsimile of an ancient photo, dog-eared and browned. It was autographed. She only caught a few words: "*To Mr. Joseph Merrick, with thanks.*" The name vaguely registered in her memory, but she couldn't place it.

Taina spotted her looking. "That's a relative of mine. A long time ago. Royalty from Denmark and then Wales. When Joseph Merrick died, the photo was given back to our family. Long story."

Gwen took her cup and stirred in some pseudo-honey. "Well, we've got an hour or so."

"True enough." Taina blew into her cup. A cloud of steam wafted into the air. "In the good old days, back in the nineteenth century or so, one of my relatives—princess Alexandra of Denmark—married Prince Albert of Wales. The two of them donated funds to build on to a hospital in London. That hospital happened to be the home of the famous Joseph Merrick, the Elephant Man."

That was it, Gwen thought. The puzzle piece fell into place. "Yes, I remember the stories."

"Stories about a freak who terrified children at carnivals? Legends about a poor, sub-human creature?" Taina said it evenly, but her voice was charged with an energy that made Gwen uneasy.

"Something like that." The familiarity of the story sent a frigid breeze slithering up Gwen's spine. She tried to shift the focus away from herself.

Taina interlaced her fingers as she sat back against the wall. In the low gravity, she was sitting at an odd angle to Gwen. "The Princess had heard of Merrick, and asked to see him while she and the Prince visited the hospital. Merrick had created a house of card stock, a small cathedral or mansion. And he had books and woven baskets. The man, deformed as he was, had many surprises. While they were there, having tea and conversation, she left that photo for Merrick. They say the visit had a profound effect on her."

"No doubt, it did on Merrick, too," Gwen ventured.

Taina smiled and nodded slowly. "Yes. I suppose you can't tell a book by its cover, can you?"

Gwen said nothing, and Taina continued. "I've watched, over the years, as Melciéna Valentine worked with the Naiads, struggled for their freedom, fought against all the prejudice and preconceptions that seem to come with human nature. Maybe that's the human part: that we rise above our nature. Don't you think?"

Gwen had stopped drinking. She had stopped breathing. She said, "I think that's part of it. And part of it is knowing where to put our energies, where to focus our love."

An alarm rang out in the cockpit. Beyond, a great vessel gleamed in the sunlight, a sleek craft unlike anything Gwen had ever seen. Taina leaned forward. "We've arrived. I think you have lots to think about on your trip home. Think fast. With our new ships, it's going to be a short voyage. Good luck, Gwen. Melciéna was right. You are a 'good egg.'"

Chapter 40
Homeward

Ten Weeks Later

Adjutant Chevalier sat at his desk, looking at the files and hoping he had not made the wrong decision. His desk monitor chimed. "Yes?"

"Your fourteen hundred is here, sir."

"Send him in, please."

A young man stepped through the door, eyes hooded beneath deep black brows. His hair had been slicked back, but Chevalier could tell it was longer than in the curriculum vitae. His pale complexion made the black pate even more striking.

"Come in, Thomas."

"Thank you, sir." Chevalier smiled as the young man shook his hand rather than saluting. Chevalier guessed that the military gesture would be alien to him. "And because of our working relationship, why don't you call me Armand?"

'Yes sir, Armand. Thank you for the appointment. The hiring." Tommy's arms seemed to be in all the wrong places. "I'm happy I got the job."

"I'm happy you got it, too. Now that you're not in a job interview, why don't you tell me more about this vision of yours."

"My vision. Hm. I suppose I got the historian bug while I was working as a librarian. I was actually what they call a page; I shelved books. Real books; not digital files. And I knew the place. Knew all the sections. And I saw how our long saga, from taming fire to leaving the home world for the worlds out there, all that was just about lost in Wentaway. I wanted to be a part of saving it, of reconstructing the past. And as soon as I heard about the Naiads, and saw how they might change the course of history, I wanted to be here, all in."

"And travel doesn't bother you?"

"I never went that far."

"Few people have. But Iapetus will be a good base for your research. We're building up the infrastructure there, because of our trade with the Naiads, and the fact that we can't make landfall on Enceladus."

"Looking forward to it, sir."

"Now, I believe you have an important appointment with someone down the hall. My secretary will accompany you. You have my numbers. Use them. I want you to think of me as your personal assistant here on Earth. And I'm looking forward to reading your Human/Naiad history."

***

Rhodes hadn't seen his nephew since Tommy was a toddler. They'd kept up some, but the call of the job had kept Carter away, both physically and electronically. Now, his late sister's son stood before him, sharply dressed in relatively appropriate civilian garb. His temporary nametag read, "Thomas Belfort."

Rhodes stood, came around his prodigious desk, and shook Tommy's hand.

"Hello, Uncle Carter."

Rhodes let go of Tommy's hand and gave him a bear hug. "Thomas, I want you to know that I had nothing to do with the hiring. You earned this, fair and square."

"Did they know we were related?"

"Chevalier is a very thorough researcher. But I was not involved, just a curious bystander. How's my brother-in-law? Keeping you in line?"

"Dad's good. Working away. Happy, I think. He keeps busy since Aunt Rosie died. I'm sorry about your sister."

"Thank you. I'm glad your father is well. When do you ship out?"

"End of the week."

'Tommy, I have something to tell you, but it needs to be kept quiet for now."

He shrugged, looking the part of the teenager. "Sure. What?"

"I've been able to arrange something for you. It's going to break a few rules, but you are in a unique position. You will be representing two races and their combined history. That's a history that's not finished, and may never be. You'll need to do the best job possible."

"Yessir."

"So I spoke with my contacts out there in the hinterlands of ice moons. There is a Naiad there who has an interest in your project. His name is Thrk."

"Thrk." Tommy tried the word, savored how strange it felt on the tongue.

"He has convinced the Naiad council to allow you a short virtual tour. Inside Atlantis West. This will necessitate a trip to the surface, which goes against all sorts of diplomatic protocols, but all parties have agreed. They are allowing you to station yourself on the surface station, set up a little office for your research. From there, you can tour the place below—remotely, of course. I just don't want you to flaunt it around here. This is hallowed ground you'll be stepping on, the inner sanctum. People would give their right…eye to go there, to actually see it. But wait to talk about it until you're at Iapetus. Everybody there will know what's up, and they'll help you connect with a special ship to take you down, and back up again when you're done. It's a beautiful little ship called the *Belvedere One*."

"Sounds exciting. To see those places where it all began."

"Yes, the things that you are going to see are breathtaking. The entire settlement is built like a giant butterfly. The spine looks like a glass cathedral. And the sculpture. Naiad sculpture is fascinating. You could do an entire chapter on that."

"Yes," Tommy smiled. "Maybe I will. Hey, what is this?"

"Come see." Rhodes ushered his nephew across the room to a low table. On it was spread a 3D printed model of a village. "We're calling this New Breckenridge."

"It's even got little trees."

Rhodes could tell Tommy's tone was a bit dismissive. "Yes, well, it's a big project. Opening is planned for late autumn of 2231. The ski slopes and snowmaking stations are already in place. You have a long road ahead, Tom, but when your travels bring you back this way, you and I can take a few days and go shooshing. Maybe we'll bring a friend of mine. She's on the Moon right now, but she's got the wanderlust. She'll be back here one of these days, and she's expressed interest in my little New Breckenridge project."

"A friend?"

"Colleague."

Tom feigned frustration. "Can you describe her?"

"She's a brilliant bioengineer. Clever. Beautiful."

Tom looked self-satisfied, as if he had uncovered the culprit in a murder mystery. "Thought so. I'm happy for you."

Rhodes thought for a moment. "I guess I'm happy for me, too. So how about it?"

"Sure, Uncle Carter. Maybe we can all shoosh together."

***

Despite the fact that she had been in the inner system for weeks, Gwen still hadn't made it home to Mac IV. She had business to wrap up in Sao Paolo, and things to see to at the office complex on the Moon's orbital hub. She hoped to have business elsewhere on Earth, so Luna was the place to hang out, at least for the time being.

Gwen found herself missing the old haunts: Mac II's high deck, Mac IV's Gallagher's, Josh Aotea's frumpy coats and thick glasses and kind Kiwi ways. But at the same time, she looked forward to the fresh flowery scents and the cool spring air of Earth. Carter Rhodes had invited her to his new ski resort, and the thought seemed to be growing on her. He, too, had proven himself a good egg. How soon she would be back to the home world was anyone's guess. She hoped it would be very soon. The lunar landscape spread before her was too gray, and she wasn't quite ready to return to the golden skies of Venus. Low in the sky, the brilliant blue Earth beckoned just next door.

Her hotel window opened onto humble greenhouse gardens. Fans moved the aromas of flower, greenery and loam into her bedroom. Beyond, through the glassy walls at the greenhouse habitat's far side, lay the broken, undulating landscape of Mare Cognitum, a dark plain peppered by white powders from the great impact of the Copernicus crater. Mare Cognitum, *the sea that has become known*. Fitting that she would end up here, after her voyage to Enceladus and back. Many things had become known to her in short order. The Naiads held the imprints of human DNA, the form of the person, but they had come into their own as something completely new. Gwen wondered what new art, what new inventions, what new philosophies would be born beneath the ices of that tiny, frigid Naiad home world. Her own journey had taken her through the namesakes of other lunar landmarks: the sea of crises, the ocean of storms, and even, ultimately, the sea of serenity. She was ready, at last, to turn from her past and face a future made new by recent events.

It was Tony who first brought home for her what kind of shape that new future would take.

"Gwen, you've been so busy caring for everybody in general that you've forgotten how to care for anyone in particular." It was true. Her amorphous love for the human race—for all living races—had come at the expense of those around her. It was time to make some changes. Time to do some growing up.

Claire still lived with her family in a humble, eighty-fifth floor condominium in Philadelphia. On a clear day, they could see the Delaware River on one side and Independence Hall on the other. Gwen had stopped counting how many years it had been since they sat down to a transmission session, let alone caught up in person. The two of them had spent far too much time keeping past pain inside themselves. But Philadelphia was entering spring, a time for new life. A time to be free. Perhaps a time to reset a few things.

Gwen hoped the number hadn't changed as she typed it into the keyboard and linked to the Terra/Luna comms line. She transmitted in audio only. The Earth/Moon distance was short enough for the delay to be manageable. She heard static, and it made her think of her new friends on Enceladus. The large and the microbial.

Claire's voice on the other end had the slight echo of the Earth/Moon connection. "Hello?"

Steeling herself, Gwen said, "I never did like long goodbyes."

After a lengthy pause, the word came. A single word. A wonderful word. "Sis?"

"*Gerthingmanchester.*" Gwen said the word tentatively. She waited. Finally, her sister's voice again.

"Secrets to share. Secrets to tell. Come home, Gwen. I may even have some sunflower seeds lying around."

And in the static of the radio, or perhaps in the greenhouse breeze drifting through the window, Gwen could hear the laughter of Melciéna Valentine.

Part II
The Science Behind the Story

Chapter 41
The Science Behind the Story

Venus: Home sweet home?

Our hero, Gwen Baré, witnesses the broad sweep of Venusian weather from her perch aboard a floating city in the clouds of Venus. She sees sulfuric acid virga, watches the winds tear away at the billowing clouds, and glimpses lightning on the horizon.

Venus weather has been charted since the first spacecraft[1] passed by in 1962. On its way to a series of Mercury flybys in 1974, Mariner 10 imaged the clouds in UV, unveiling the dramatic winds there. The Soviet Union's Venera 9[2] and 10 became the first Venus orbiters, arriving in October 1975. They carried imaging systems and other experiments. Each carried a lander which successfully returned panoramas of the sweltering surface. The Pioneer Venus Orbiter mapped clouds and wind from 1978 to 1992, and created the first global radar map of the planet[3]. The European Space Agency's sophisticated Venus Express was the first long-term craft dedicated to weather observation at Venus. Venus Express was first to confirm the existence of lightning. Up until its discovery, spacecraft had returned only circumstantial evidence,[4] and researchers were skeptical, as the only lightning found in the solar system was associated with water clouds. But over the course of a three and a half year period, Venus Express demonstrated that clouds of sulfuric acid could also develop electrical storms. Venusian lightning may, in fact, be more widespread than it is on the Earth. Lightning seems to be more common on the daylit side, and is more prevalent at the equator than at higher latitudes. As of this writing, Japan's

---

[1] Mariner 2 flew within 35,000 km of the planet and was first to record Venus' high temperatures. The craft estimated a temperature of 270°C ± 50°.
[2] Venera 9 also carried the first successful Venus lander.
[3] Other radar mappers included the Soviet Venera 15 and 16 and NASA's Magellan.
[4] The Soviet landers Venera 9 (1975) and Venera 12 (1978) returned data indicating lightning or thunder, but the data was inconclusive.

Fig. 41.1 A surface view of Venus, taken by the Soviet Venera 13 lander, provides a hint of the blistering desolation beneath the Venusian clouds. (Institute of Space Research image reprocessed by Don Mitchell and Michael Carroll)

Akatsuki orbiter continues to study cloud physics, climate and global atmospheric dynamics of the hothouse world.

The surface conditions on Venus are drastic enough to challenge today's most advanced technologies. The Venusian 'garden level' simmers at 462°C, comparable to a blast furnace. Air density is 90 times that of the Earth at sea level. A walk on Venus would feel like a walk on the bottom of a swimming pool. The rocky surface of the planet is essentially basaltic, consisting primarily of ancient, weathered lava flows.

The prospect of a long-term inhabited facility on the Venusian surface is daunting. How difficult would it be to deploy a science outpost, or even a small village, in the skies of our nearest world? Super-pressure balloon probes have been proposed for both Venus and Mars, and Montgolfier-style ballooning is a possibility on Titan, suggests another study.[5] But NASA Glenn researcher Geoffrey Landis has something bigger in mind. Landis carried out a NASA-funded study[6] to see what it would take to set up a balloon-supported outpost in the clouds of Venus. The floating station would cruise at an altitude of about 50 kilometers, where Venus' atmospheric pressure is about 1 bar—equivalent to Earth at sea level—and the air is at

---

[5] A Review of Balloon Concepts for Titan by Ralph Lorenz, JBIS, Vol. 61, pp. 2-13, 2008

[6] Landis, Geoffrey, Colonization of Venus, Space Technology & Applications International Forum, Albuquerque, NM, February 2-6, 2003.

room temperature. Any humans living in those conditions would need only light clothing to protect them from the acidic air, and something akin to an oxygen mask. The Landis report points out that Venus is rich in resources necessary for life: carbon, hydrogen, oxygen, nitrogen and sulfur. Its dense atmosphere provides protection from cosmic radiation, in contrast to environments on Mars, the Moon, or asteroids, other targets for human exploration and settlement. Venusian gravity is 90% that of Earth, so the long-term physical problems that may arise on Mars or the Moon are not an issue there. Additionally, solar energy is abundant above the cloud tops, and wind energy in Venus' breezy weather would also be abundant. The planet affords another benefit to would-be sky realtors: because of the density of Venus' carbon dioxide atmosphere, a mix of breathable air serves as a lifting gas, about half as buoyant as helium is in Earth's atmosphere. A 400-meter-radius balloon—about the size of a small sports arena—can lift 350,000 tons, a mass equivalent to the Empire State Building.

Because of the global winds in the middle cloud deck of Venus, the aerial colony would not remain stationary, but would drift, continually circling the globe. Observers could constantly study new ground, and relay stations could be deployed to keep in contact with surface telerobotics. Sulfuric acid droplets would constantly batter the station, but plenty of acid-resistant technologies are already available. In the case of our story, these floating villages could also be stationary, utilizing powered propellers to remain over one site or taking advantage of Venusian winds to relocate to other viewpoints. The NASA Glenn study concludes, "In short, the atmosphere of Venus is the most earthlike environment in the solar system. Although humans cannot breathe the atmosphere, pressure vessels are not required to maintain one atmosphere of habitat pressure, and pressure suits are not required for humans outside the habitat… in the long term, permanent settlements could be made in the form of cities designed to float at about fifty kilometer altitude in the atmosphere of Venus." Despite the shortcomings of sulfuric acid clouds and excess carbon dioxide, Venus is a reasonable target for future aerial exploration, and perhaps even high-altitude settlement.[7]

Enceladus: where the action is.

Most of our plot takes place on the ocean floor of Saturn's exotic ice moon Enceladus. Enceladus was made famous by the Cassini orbiter when the craft discovered active jets of water vapor streaming from vents in the South Pole. The tiny ice moon was thought to be geologically dead, like its same-sized sibling Mimas, a cratered, inert ball of ice. But Enceladus orbits within Saturn's E-ring, and many researchers suggested that the moon was somehow feeding the material in the ring itself. They were right. A series of parallel canyons, called Tiger Stripes, play host to nearly 100 jets

---

[7] For more on Venus colonies in the clouds, see *Drifting on Alien Winds: Exploring the Skies and Weather of Other Worlds* by Michael Carroll (Springer 2011)

of ice particles, some thundering 400 km high. The particles that fall back to the surface are rich in salt, and have been in contact with the stony seafloor. Finer, more pure water particles make it into orbit, recharging the E-ring.

Enceladus' fountains issue from vents estimated to be just 9 meters across. Eruptions occur in long lines within the Tiger Stripe canyons, with each fissure extending some 130 km. Some models indicate that the crust at the southern pole is thinner than at other sites around the globe, measuring as thin as 5 km. Researchers are still grappling with the mystery of how—and why— the subsurface waters make their way to the eruptive sites in the vacuum of space. Tidal flexing of the ice is likely the primary driver the plumes. One hint implicating tidal forces is that the brightness of the plumes changes with orbital location. When Enceladus is at its most distant location in its oblong orbit, the most activity occurs. At its closest approach to Saturn, only a third as much material escapes.

Several theories have been put forth to explain the plumbing of the jets.[8] One concept posits a carbonated sea. As water rises through fractures, dissolved CO_2 comes out of solution, forming bubbles. A second theory holds that water vapor and gases accelerate as they flow through nozzle-like channels in the Tiger Stripes. A third concept has more to do with tidal stresses, which might cyclically force fractures opened and closed, heating the water inside.

It was Enceladus' geyser-like activity that first led researchers to wonder just how much water lurked under the ice crust. At first, models suggested a lens-shaped ocean sloshing beneath the ice at the South Pole. But as Cassini made several close passes, investigators were able to study the changes in its radio signal, which indicated subtle variations in the spacecraft's velocity. These changes are due to deviations in the gravity, which is affected by the structure of Enceladus. The measurements were consistent with a sea 30 to 40 km below the surface of the South Pole, and extending up to 50° south latitude. Careful study of surface images proved that the entire crust of the moon was librating, or wobbling. This means that the crust of Enceladus is decoupled from the moon's core, unattached and floating on a global sea.

The Planetary Science Institute's Candice Hansen-Koharcheck is a leading expert on the subject. She says that according to the best models[9], "the icy crust shell is 21-26 km thick, then comes the ocean at 26-31 km thick, then the core. The numbers are different at the south pole – the crust is thinner and the ocean is deeper, but it is all very model-dependent." Whatever is causing them, the flittering jets of water blast out prodigious amounts. Enceladus erupts roughly 200 kilograms (440 pounds) of water into space every second.[10]

---

[8] The following articles all appear in *Science* **311** (2006): C. Porco et al., 1393; J. Spencer et al., 1401; M. Dough- erty et al., 1406; F. Spahn et al., 1416; J. H. Waite Jr
  et al., 1419; C. Hansen et al., 1422.

[9] For more on this subject, see **The Gravity Field and Interior Structure of Enceladus**, *Science*, April 4, 2014, by L. Iess, et al, and **Enceladus's measured physical libration requires a global subsurface ocean**, *Icarus*, Volume 264, January 15, 2016 by P.C. Thomas et al.

[10] **Enceladus' water vapor plume**, *Science* March 10, 2006, Hansen CJ, Exposito L, et al

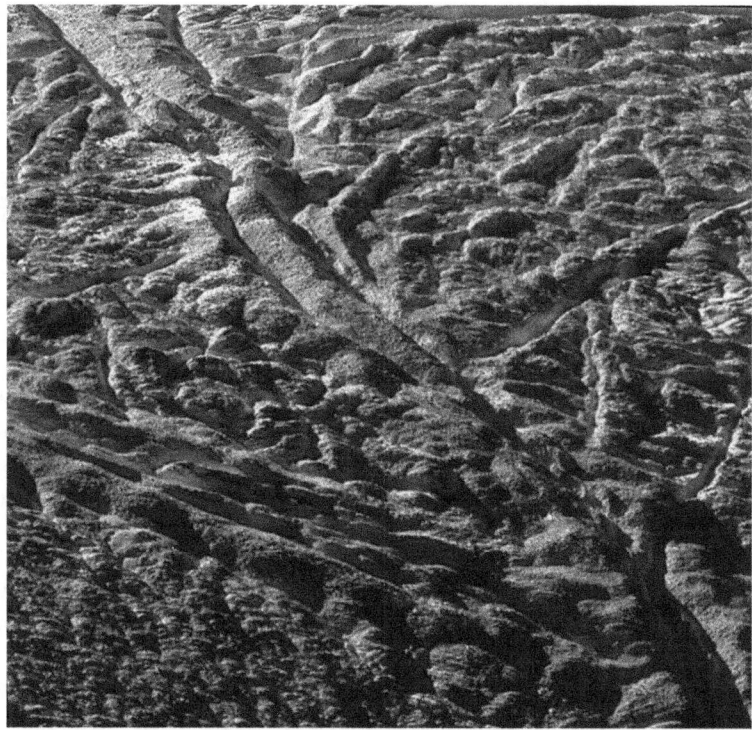

Fig. 41.2 Imaged by the Saturn-orbiting Cassini spacecraft, this photo mosaic is one of the most detailed views of the volcanic provinces in Enceladus' southern hemisphere. Taken at an average range of 124 km (77 miles), each pixel captures an area ~15 meters square. (NASA/JPL/Cassini Imaging Team)

But what about life? Our characters come to realize that the Enceladan sea is full of it, but is life on the tiny cryogenic moon really possible? Is it likely?

The instruments aboard the Cassini spacecraft had several opportunities to directly sample the plumes, flying through them and skimming the surface as close as 49 km (the craft had earlier made an even closer pass at the equator, with a low altitude buzz of just 25 km). In all, the spacecraft reconnoitered the moon 23 times in the course of its 13–year Saturn mission. Its Ion Neutral Mass Spectrometer sniffed a smorgasbord of constituents in addition to water, including carbon dioxide, ammonia, methane, and several other hydrocarbons, the building blocks of complex, life-related organics like amino and nucleic acids.. But Cassini revealed an even more significant find for those searching for extraterrestrial life: the likelihood of hydrothermal vents on the seafloor. Cassini's Cosmic Dust Analyzer detected microscopic particles of silicate, demonstrating that the moon's ocean is in contact with the seafloor, where minerals necessary for life abound. These silica particles form in very hot water, like that found in the plumes of submarine volcanoes on Earth. Additionally, the methane in the plumes can exist for only a short period before it is locked in rock or ice, and the most likely source of its replenishment is hydrothermal seafloor activity. A final item of evidence that researchers cite for the existence of seafloor volcanism is the presence of hydrogen in the plumes. A likely producer of this hydrogen is called serpentinization of rock, a process in which water is chemically bound to lava, and the rock is oxidized, converting it into serpentinite. This process could provide the chemistry needed for deep-sea life at the rocky ocean floor of an ice moon. On Earth, where we have observed hydrothermal activity up close, life thrives in the deep.

Enceladus is the whitest object in the Solar System,[11] says Jet Propulsion Laboratory researcher Morgan Cable. "On Enceladus, the plume is sending out so much material that the older ice is covered up quickly by fresh material, which is why it's so white and bright."

Still, Cable says explorers may see many color subtleties within the Enceladus landscape.[12] "Glacial ice on Earth is blue due to scattering of blue light and absorption of red; I imagine the same thing would happen if Enceladus ice was exposed to sunlight in just the right way as well. We just haven't been close enough to see this perhaps. Space weathering can cause surfaces to darken over time. We think, for icy bodies, the darkening might be due to the formation of complex organics ('tholins') from simple molecules like nitrogen and methane. But we don't really see this on Enceladus (we do on other moons of Saturn, like Iapetus)." One thing that might color the ices on Enceladus is salts changing color due to radiation processing. Some salts, like sodium chloride, change color when exposed to radiation like the Sun's UV light.[13] Sodium chloride turns yellow-orange, and another salt, potassium chloride, turns purple.[14]

Energy from Microbes: a brainstorming session with microbiologist John Coates

Our hero, Gwen Baré, specializes in getting power from microbial colonies. Today, this technology is in its infancy, but strides are being made by a host of researchers. One focus is the microbial fuel cell (MFC), also known as the bioreactor. Its operation is similar to that of a battery, says Dr. John Coates, director of the Energy Biosciences Institute and a senior scientist at the Lawrence Berkeley National Laboratory. "In an ordinary battery, you have an electron-dense substance on one

---

[11] For further research on surface properties and colors, see **Icy Saturnian satellites: Disk-integrated UV-IR characteristics and links to exogenic processes,** by Amanda Hendrix, et al, *Icarus* Volume 300, January 15, 2018

[12] For more on human exploration of Enceladus, see *Living Among Giants: Exploring and Settling the Outer Solar System* by Michael Carroll (Springer 2015)

[13] https://agupubs.onlinelibrary.wiley.com/doi/full/10.1002/2015GL063559

[14] Berzina, B. (1998). "Formation of self-trapped excitons through stimulated recombination of radiation-induced primary defects in alkali halides". *Journal of Luminescence*. 76-77: 389

Energy from Microbes: a brainstorming session with microbiologist John Coates

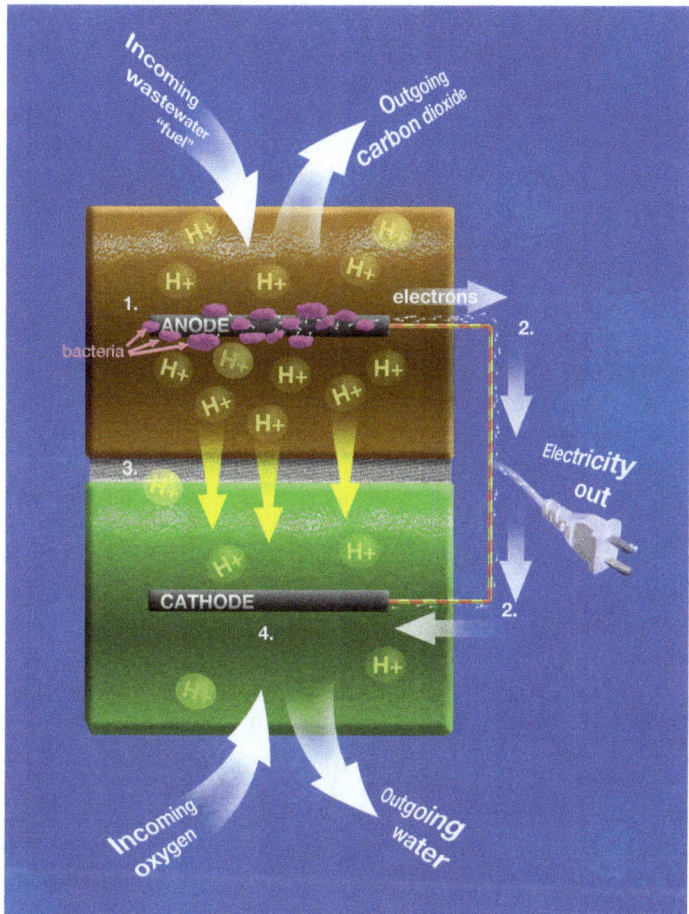

Fig. 41.3 Diagram of microbial fuel cell: bacteria break down food and organic waste, turning normally discarded materials into energy and water. One day, microbial fuel cells may produce higher energy output, as in our story, but today's bioreactors produce a gradual flow of low current, along with byproducts like hydrogen for fuel cells and fresh water. 1. The bacteria on the anode break down organic matter, releasing electrons and hydrogen ions (H+). The electrons flow from the bacteria onto the anode. 2. The electrons migrate up from the anode. A wire directs them onto the cathode. While flowing through this wire, they generate an electrical current that can be used to perform work. This is the point in the process where useful electricity is generated. 3. The abundant hydrogen ions flow through a semi-permeable membrane to the cathode, where there are fewer hydrogen ions. 4. The electrons from the cathode combine with dissolved oxygen and the hydrogen ions to form pure water (H_2O). (Diagram by Michael Carroll)

side of the battery and an electron-poor substance on the other side. Electrons try to move from the dense substrate to the electron poor material." Electricity is the movement of electrons. This movement is simply nature trying to keep things in balance. Weather is the product of a planet's attempt to even out temperatures by mixing cool and warm air. Currents in the ocean come from the Earth's rotation, but also from temperature differentials that move seawater. Coates likens nature's balancing operation to a person carrying water buckets. "If you're carrying two buckets on a stick across your shoulders and one bucket is full of water and the other one is empty, it's going to be very unbalanced, so what you want to do is balance it by moving some of the water from one bucket to the other. Everything in nature wants to move toward equilibrium. With that in mind, what you need to make a system like an MFC work is a system that's oxidized (oxygen rich) next to a system that's reduced (electron rich)."

Engineers took advantage of this fundamental electrical principle in the human space program, making use of fuel cells that combined hydrogen and oxygen. Hydrogen is electron dense, while oxygen is electron poor. Electrons will move from the hydrogen to the oxygen, and the end products are electricity and water.

Newer technology[15] [16] enables engineers to leverage this electron movement in the microbial world. The natural environment contains many such electron-dense and electron-poor substances. Microorganisms are capable of utilizing oxygen in the same way we do: they breathe it in and reduce their oxygen by adding an electron onto it, creating water vapor. In doing so, they move electrons. But microorganisms can use other compounds, such as sulfates and nitrates. "When you have a high density of microorganisms growing in the environment, they can make that environment energy rich," Coates explains. "If you connect an electrical wire between an energy rich environment and an energy poor environment, such as a water column that has oxygen in it, now the electrons will move back from the energy-rich environment to the oxygen in the energy-poor environment. Then we stick our light bulb or whatever we want along that wire, and that creates resistance and we put it to work."

While microbe generation of electricity can be affected by pH or temperature, researchers have successfully used microbes to power offshore ocean sensors. In part because of microbial activity, ocean water is oxygenated, while sediments in the ocean floor are anoxic (lacking in oxygen). In utilizing that difference in oxygen levels, engineers can use the ocean itself as a global battery.

In a promising new application, bioreactors have successfully generated energy while purifying water.[17] Microorganisms pass electrons onto an electrical wire by metabolizing waste material, like the sewage and organic waste that colonists in our undersea Enceladan outpost produce. Bacteria can degrade the waste, converting it to carbon dioxide. They generate electrons in that reaction, and pass those electrons on to an electrode. The process sets up an energy-rich electrode. Microorganisms can also be stationed at the other end of the electrode that takes electrons off: those organisms pass the electrons on to something like sulfates that might be present in

---

[15] **Microbial Fuel Cells: Methodology and Technology** by Bruce E. Logan, Bert Hamelers, René Rozendal, Uwe Schröder, Jürg Keller, Stefano Freguia, Peter Aelterman, Willy Verstraete, and Korneel Rabaey. *Environmental Science & Technology* **2006** Vol. *40*, pp5181-5192

[16] **Microbial fuel cell as new technology for bioelectricity generation: A review**
by Mostafa Rahimnejad, Arash Adhami, Soheil Darvari, Alireza Zirepour, Sang-Eun Oh. *Alexandria Engineering Journal* **2015** Vol. *54*, pp745–756

[17] See **Production of electricity during wastewater treatment using a single chamber microbial fuel cell**, by Hong Liu, Ramanathan Ramnarayanan, and Bruce E. Logan, *Environmental Science Technology*, February 21, 2004

seawater. The passing of electrons creates electrical current that can be used by anything from terrestrial industries to Enceladus settlements.

One organism that John Coates' team is working with is called *Geobacter sulfurreducems*.[18] The *Geobacter* organism carries a unique set of structures to shed its electrons, Coates says. "To get an electron outside of the cell, the organism has to have some kind of electron carrier. We take oxygen into our cells and we add the electron onto the oxygen inside the cell. What *Geobacter* does is it produces these nanowire filaments that are electrically conductive. These carry the electrons outside of the cell, allowing the organism to pass them onto a solid surface. So the organism can attach itself to a solid surface like an electrode, and can easily allow the electrons to flow down through these connections that it makes, onto the surface of the electrode."

Microbial fuel cells can generate electricity, but the problem is the amount that the microbes can produce per square meter of electrode surface. "The problem is that the microbes are not doing this for our benefit," says Coates. "They do it for their benefit. They're going to take a lot of that energy for growth. The reality is that microbial fuel cells will generate very low voltage. MFCs are limited by the number of electrons that can be moved from the microbes at any given time. The surface of the electrode can only fit so many cells before it becomes saturated. You can't get a huge amount of electrons moving at once. Power yields are quite low. But where microbes can alternatively use waste to make methane gas, they can do that very efficiently. They can make copious quantities of methane. And methane gas can be stored as an energy source. Potentially, that methane could be used in a chemical fuel cell and be converted back into electricity." In fact, researchers at Michigan State University[19] are successfully taking this two-pronged approach. Using two microbes, their process imitates natural waste digestion while generating 20 times more energy than existing processes by creating ethanol and hydrogen for chemically based fuel cells.

For our terrestrial, Enceladan and Venusian colonists, Coates envisions a mixture of power coming from the bioreactors. The first order of energy from an MFC, direct power, can be used for small-scale requirements. MFCs can recharge phones or run sensors. "Right now, microbial fuel cells can't generate those huge amounts of power. People have looked at them as long-term trickle chargers, where you can recharge a capacitor or battery over the long term and then use that battery in the short term. You can make methane gas or hydrogen, and that can be rapidly converted to electricity in large amounts. But you generate two types of energy at once. The microbial fuel cell might be able to run all the lighting from low-power LED's, and you can recharge continuously. The microbes give us free power; they just keep

---

[18] **Structural optimization of contact electrodes in microbial fuel cells for current density enhancements**, by Shogo Inoue, John D. Coates, et al, *Sensors and Actuators A: Physical* (Elsevier), Volume 177, April 2012.

[19] For more on their work, see **Electricity generation in microbial fuel cells using neutral red as an electronophore**, by Doo Hyun Park and J. Gregory Zeikus, *Applied and Environmental Microbiology*, accepted January 7, 2000

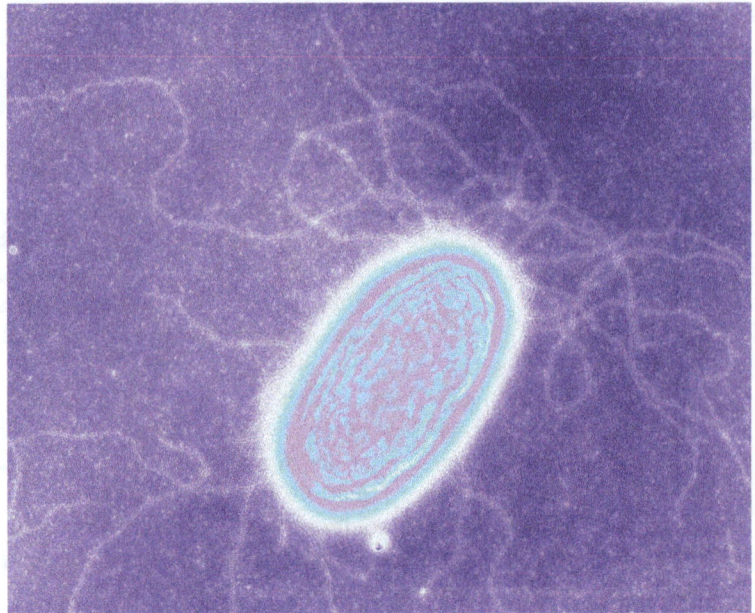

Fig. 41.4 Geobacter metallireducens bacterium, colored transmission electron micrograph. This anaerobic bacteria oxidizes organic compounds to form carbon dioxide, using iron oxide or other metals as an electron acceptor. Geobacter can be used within a microbial fuel cell and generate electricity by oxidizing waste organic matter and transferring surplus electrons directly to an electrode. Geobacter grow long filaments known as pili that are electrically conductive. Geobacter was discovered in 1987 by Derek Lovley, Professor of Microbiology at the University of Massachusetts, USA. (Derek Lovley/Science Photo Library)

doing their thing. You're just adding a wire into the mix of flowing power. It's just a question of the best way to utilize it." And with advances in technology, bioreactors may one day be scaled up to power entire cities, on Earth and beyond.

Broadcasting Bacteria[20]

Do microbes transmit radio waves? The question has engendered some controversial research and heated discussion.[21] Nobel recipient Luc Montagnier, the virologist who first linked HIV to the AIDS disease, suggested in 2009 that microbes might communicate amongst themselves using some form of radio signals.[22]

---

[20] **Electromicrobiology** Derek R. Lovley. *Annual Review of Microbiology* **2012** Vol *66*, 391-409

[21] For a sample of recent research, see **Electromagnetic homeostasis and the role of low-amplitude electromagnetic fields on life organization** by Antonella De Ninna and Massimo Pregolato, *Journal of Electromagnetic Biology and Medicine*, Volume 36, issue 2, 2017

[22] **Electromagnetic signals are produced by aqueous nanostructures derived from bacterial DNA sequences**, by Luc Montagnier, et al, *Interdisciplinary Science,* June 2009

Montagnier's proposal is controversial on several counts. One of the criticisms leveled at the concept was that bacteria have no known mechanism for generating radio waves. Another criticism specific to the 2009 study is that thus far, no investigations have shown any bacterial response to radio waves. Another question that arises is what kind of message or data could be transmitted and received by simple organisms. But cells do often work in concert (for example, the muscle cells of a beating heart or the islets of Langerhans in the pancreas). Our novel posits a sort of hive mind arising in such group communication.

Cells utilize electromagnetic waves at higher frequencies to communicate, as well as to send and store energy. Some bacteria communicate via nanowires (like Geobacter, mentioned above). If cells can also generate radio waves, some microbiologists reason, they may well exploit this path to communicate across microscopic communities. [23, 24]

In our adventure, humans, Naiads, and microbes inhabit the oceans of Enceladus. Nature exhibits instances of what biologists call "Swarming Intelligence." The phenomenon involves a population where individuals interact with each other in concert. In the natural environment, swarming individuals follow simple actions or patterns interlaced with those actions of other individuals. As in our Enceladan plankton, even when there is no central "brain" dictating how individuals should act, interactions between members of a group lead to the rise of "intelligent" behavior, different from behaviors of individuals. Examples of Swarming Intelligence include the flocking of starlings and other birds, ant colonies, mammal herding (i.e. buffalo, Wildebeests or domestic cattle), and the schooling of fish. Even microbes have demonstrated swarming or group behavior. Some bacteria carry out complex behaviors in single cells, and collaboration in populations of similar or even different cells. The collective actions seem to be controlled by chemical signals that trigger changes in cells, both physiological and behavioral. These group behaviors can influence the structure of entire colonies. One example is protozoans, which exhibit the ability to organize themselves when they detect changes in their environment. Even bacteria, which behave in primitive and reflexive ways as isolated cells, sometimes operate in more sophisticated ways as a large population.

Biofilms offer another instance of microbial cooperation. Biofilms occur when microbes bind together to create a thin, slimy film on a surface. Biofilms formed by Bacillus subtilis synchronize the growth of the entire film by generating electrical impulses. These signals assure that the innermost cells of the biofilm do not starve. Under environmental stress, when food is short or unavailable, other microbial colonies organize themselves in such a way so as to maximize the availability of nutrients.

---

[23] For more on the subject, see **Low Frequency electromagnetic waves as a supplemental energy source to sustain microbial growth?** By Victor A. Gusev and Dirk Schulze-Makuch, Naturwissenschaften, March 2005, Volume 92, Issue 3, pp 115-120

[24] **Microbial Communication, cooperation and cheating: quorum sensing drives the evolution of cooperation in bacteria** by Tamas Cazaran and Rolf F. Hoekstra, *Plos One*, August 17, 2009

Fig. 41.5 Schools of fish demonstrate swarming intelligence. (Image courtesy Seacology/ Wikipedia: https://commons.wikimedia.org/wiki/File:School_jacks_klein.JPG)

Individual cells of slime molds (like myxobacteria) work together to produce complex structures or move as multicelled groups (also called "wolf packs").

Populations of bacteria use quorum sensing—behavior affected by population density—to judge their own concentrations and change their behaviors accordingly. Some social insects (bees, ants) use quorum sensing to determine where to nest. Quorum sensing can function as a decision-making process in any decentralized system, such as computers and robotics. For it to function in a group setting, the individuals must be able to sense the number of other components they interact with, and they must adopt a standard response once the group reaches a threshold number of individuals.

These examples in nature provide our story with a model for the global plankton of Enceladus.

Incoming!

The backstory of *Lords of the Ice Moons* involves the impact of an asteroid, a world-changing event whimsically referred to as *Wentaway*. Asteroid impacts have left their mark across the face of our own world in many places. In Winslow, Arizona, a 110,000-ton mountain of metal vaporized a Pleistocene landscape. Its 2.5-megaton blast left the Barringer Meteor Crater, a dramatic bowl spanning 1.2 kilometers

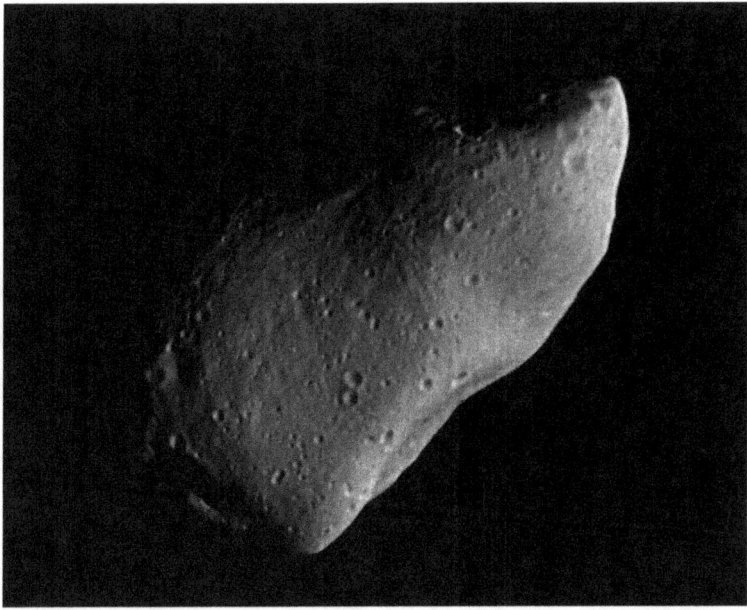

Fig. 41.6 The asteroid 951 Gaspra is roughly the size of the impactor in our story. The Jupiter-bound spacecraft Galileo snapped this shot of Gaspra from a distance of 5,300 km (3,300 miles). Objects as small as 54 meters can be seen on the surface. Like most asteroids, Gaspra is irregular, with dimensions of 19 x 12 x 11 km (12 x 7.5 x 7 miles). (Photo courtesy NASA/JPL)

across and 180 meters deep. The crater's hummocky rim rises 30 to 60 m above Arizona's rolling plains. The iron-nickel meteor that created it may have been as large as 60 m in diameter, as far across as three railroad boxcars.

In more recent times, Earth has been visited by two cosmic intruders that exploded on their way to the ground. On June 30, 1908, an incoming intruder exploded above the forests of Siberia. The impactor didn't make it to the ground. The object was likely a comet nucleus, filled with frozen gases, which detonated 6 to 12 km above the Tunguska River in a nearly uninhabited area. The airburst leveled 80 million trees and killed wildlife across an area of 2150 km^2.

A similar event took place in more modern times, near the Russian city of Chelyabinsk, a city of 1.3 million on the border of Siberia. On the morning of February 15, 2013, shortly after dawn, a bright fireball blazed across the sky from the southeast. As with the Tunguska event, the object exploded before it got to the ground, disintegrating at a location some 40 km south of the city at an altitude of just over 23 km. Fragments of the bollide, or exploding meteor, showered an area surrounding Lake Chebarkul, where some have been recovered. Many drivers in Chelyabinsk equip their cars with dashboard-mounted cameras, and some recorded the meteor's spectacular descent. Thousands of people witnessed its dazzling flash. Bystanders and office workers rushed to windows to see the source of light and the glowing trail that stretched across the sky. The shock wave, moving at the speed of

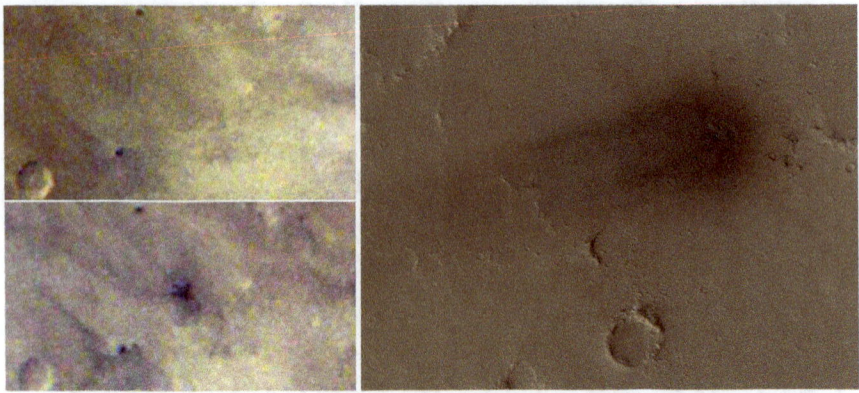

Fig. 41.7 Fresh Mars craters: (left) before and after views of an impact on Mars; (right) a fresh crater drapes dark debris across the Martian plains. (images courtesy NASA/JPL/MSSS)

sound, reached those people with a delay of nearly two minutes. Many were gazing through their windows when the blast shattered the glass. In the end, 1491 people reported serious injuries, from temporary blindness to mild ultraviolet burns (akin to sunburn) to lacerations from flying glass. 7200 homes, factories and warehouses were damaged by the meteor's shockwave. Inhabitants of Chelyabinsk reported sulfur or gunpowder smells beginning an hour after the meteor's blast, lasting into the evening.

The Chelyabinsk incident raised worldwide alarm, according to former Apollo astronaut Russell Schweickart. Schweickart is cofounder of the B612 organization, a nonprofit association dedicated to the detection and mapping of Near Earth Asteroids, as well as the protection of Earth from these cosmic vagabonds. "The most important thing that we learned from Chelyabinsk was that an object of that size—about 18 meters in diameter—could do considerable damage on the ground. It didn't kill anybody, but it certainly wasn't far from doing that. The town of Chelyabinsk was some 40 kilometers to the side of the ground track, and it came in at a very shallow angle so that the overall energy was deposited over a fairly large area. What that said is objects of a smaller size than we thought can do pretty serious damage." Current estimates suggest that a similarly sized impact takes place somewhere on Earth once every 1600 years on average, and about once each 6000 years on dry land.

Even larger impacts may be in our future, if history is any indicator. At the end of the dinosaur era (the Mesozoic Period), a massive asteroid slammed into the ocean off the coast of what is now the Yucatan Peninsula. Compared to the Chelyabinsk and Barringer meteors, Chicxulub was a terrifying apocalypse. The devastating explosion equaled 50 trillion Hiroshima atomic bombs. The shock pulverized the asteroid into fine grains that rushed back up into the atmosphere and into space. A globe-encircling shroud of soot brought on an instantaneous global winter. Shocked quartz crystals, indicative of great pressures, would have rained from the sky. Fragments of shocked quartz have been found in geologic formations associated with Chicxulub. Adding to the mayhem, the heat of the impact and its fallout would have triggered worldwide flash fires. The smoke would have further contributed to the atmospheric haze, blocking the Sun across the entire Earth. Temperatures dropped, photosynthesis virtually ground to a halt, and the food chain collapsed, beginning with the preeminent carnivores like Tyrannosaurus, and trickling rapidly down in a disastrous ecological cascade.

What effects might we expect from an asteroid the size of the one in our story? Dan Durda, an asteroid expert at the Southwest Research Institute in Boulder, Colorado, comments, "We're talking about an asteroid on the order of a half kilometer or a bit larger. There is enough chlorine and bromine in an ordinary chondrite composition object of that size to catalyze the complete destruction of the ozone layer. I don't know what the recovery timescale for that would be afterwards, but at least for a while whatever plant recovery is going on from all the other insults is going to have to deal with a lot of sunburn as well."

The aftermath of such a catastrophe will require international recovery work, Durda suggests. "There's all the extra socio/economic/political turmoil going on as well—think New Orleans Katrina across the affected region for a while. I don't think it'd be Mad Max, but it wouldn't be Mayberry either for sure..."

In *Lords of the Ice Moons*, scientists failed to save the Earth from the incoming asteroid. Changing an asteroid's path or destroying a threatening cosmic rock will be a difficult prospect, as our scientists in *Lords of the Ice Moons* found. Many solutions—from breaking up the asteroid with a nuclear device to strapping an engine on it to change its course—have been proposed, but they all require time. Fortunately, statistics are with us. A direct hit is highly unlikely. A more reasonable scenario is that an asteroid will pass near enough to the planet to be captured by its gravity. The region in which an asteroid must pass to be harnessed by Earth is called the keyhole. Should a space rock pass through this imaginary window, it will enter into a looping orbit that brings it back to the Earth, this time to impact. The time that it takes to travel around this last orbit will be a critical period for us to do something about it.

Before we can save ourselves from an asteroid, we have to find it. Astronomers today are mounting searches for asteroids that regularly cross the Earth's path, and proposals are on the table to send up space-borne observatories to further chart potential hazards. NASA and other space agencies are studying ways of deflecting asteroids. Even the United Nations has a task force dedicated to the threat, and to global education on the subject.

The impact at the end of the Mesozoic seems a long time ago. Even Tunguska and Chelyabinsk seem remote in the grand parade of history. How likely are impacts today? Some 48 tons of meteoritic material, from dust and sand-grain sized to boulders, fall on Earth each day. The largest to fall each day, on average, is about 40 cm. A 4-meter-diameter meteor will make landfall once each year, on average, and a 20-meter rock will hit terra firma once each century. Most hit the ocean or fall in uninhabited regions.

We find one indicator of how many meteoroids are floating around out there by observing the planet next door, Mars. The Martian atmosphere is far thinner than

Earth's, equivalent to terrestrial air pressure at 100,000 feet altitude. Consequently, it is easier to see the effects and frequency of incoming space debris. With advanced orbital imaging systems, researchers now estimate that Mars suffers about 200 impacts each year that result in craters at least 12.8 feet across. Many are clusters of craters. These impacts teach us that the interplanetary space around us is an active place fraught with debris that can change the course of history. It has in the past, and will in the future.

Naiad Brain Power

Researchers have long struggled with establishing a metric for intelligence. It is not an easy task. Since humans are the ones making the determination, human brain-power is the default scale. For example, the ultimate test of machine intelligence (i.e. has a computer become sentient?) is called the Turing Test. Simply put, a computer passes the Turing test when it is able to exhibit behavior indistinguishable from human behavior.

Fig. 41.8 The elephant's brain-to-body weight ratio is similar to that of the human, but the density of neurons in its cerebral cortex differs considerably. (Photo by Alexandra Carroll)

Computer smarts are referred to as "artificial Intelligence," but biological, non-manufactured intelligence is another case entirely. In the animal kingdom, we have come across many kinds of intelligence. What yardstick do we use to measure the intelligence of another species? Often, researchers compare behavior in order to judge. Crows and Ravens famously use tools, as do some primates. But the use of tools is only one indicator. What about the cetaceans, whose large brains can accommodate echolocation and complex three-dimensional thinking? Some research suggests that dolphins can distinguish numbers, and have complex problem-solving capabilities and inventive behaviors.

In the search for intelligence markers, researchers sometimes turn to the ratio of brain to body weight. The average adult human brain weighs 1.5 kilos. An elephant brain typically tips the scale at 4.5, while the brain of a Grey Whale may weigh 9 kilos. So, as our character Melciéna Valentine pointed out, elephant brains weigh in at three times the weight of human brains, and are of comparable mass in relation to their large bodies. Dolphins and other cetaceans also have a low brain/body weight ratio. Clearly, this is not a foolproof test. As Valentine pointed out, the structure of the brain is far more important. In particular, the number of neurons in the cerebral cortex—the center of logic, abstract thinking, and reasoning—is what sets human brains apart.

Laboratory techniques by neuroscientist Suzana Herculano-Houzel of Vanderbilt University enabled her to essentially "count" the number of neurons in the brains of various species, and to compare neuron density in different sections of the brain. Herculano-Houzel used detergent to dissolve the material that binds brain matter together, leaving free-floating neurons in what she refers to as a "brain soup." Suzana Herculano-Houzel found that it is not the size of brain, but rather the density of the neurons, that determines intelligence.

Herculano-Houzel's research indicates that the human brain contains 86 billion neurons. Of those, some 16 billion reside in the cerebral cortex. This ratio may be the most critical. And the human brain costs more energy to operate than the brains of other species. Reinforcing its importance is the fact that the human brain amounts to only 2% of overall body weight, but it utilizes 25% of the body's energy.

In our story, Valentine clones Naiads to be more intelligent than their original designs called for. What kind of neuron density would a sea-going Naiad have?

Trying to determine the subtleties of intelligence in biological forms is a slippery slope, fraught with human prejudice and subjective evaluations. In the present and in the future, these can result in the worst kind of reaction: xenophobia on an interspecies scale.

GPSR Compliance

The European Union's (EU) General Product Safety Regulation (GPSR) is a set of rules that requires consumer products to be safe and our obligations to ensure this.

If you have any concerns about our products, you can contact us on

ProductSafety@springernature.com

In case Publisher is established outside the EU, the EU authorized representative is:

Springer Nature Customer Service Center GmbH
Europaplatz 3
69115 Heidelberg, Germany

www.ingramcontent.com/pod-product-compliance
Lightning Source LLC
LaVergne TN
LVHW010340260326
834688LV00036B/804